Plain-English Study Guide for the FCC Amateur Radio General Class License

2019-2023 Edition

Richard P. Clem, WØIS

Illustrated by Yippy G. Clem, KCØOIA

Copyright © 2019 Richard P. Clem and Yippy G. Clem

All rights reserved.

ISBN-13: **9781475216615**

Richard P. Clem
PO Box 14957
Minneapolis, MN 55414

CONTENTS

Introduction		i
Chapter 1:	Impedance and Reactance	1
Chapter 2:	Transformers	12
Chapter 3:	Calculating Power	17
Chapter 4:	Combining Components	21
Chapter 5:	DeciBels	29
Chapter 6:	Other Components	35
Chapter 7:	Radio Propagation	48
Chapter 8:	FCC Rules	65
Chapter 9:	Operating Procedures	100
Chapter 10:	Abbreviations and Q-Signals	105
Chapter 11:	Equipment and its Operation	111
Chapter 12:	Single Sideband	130
Chapter 13:	Frequency Modulation	140
Chapter 14:	Digital Modes	142
Chapter 15:	Mobile Operation	158
Chapter 16:	Antennas	162
Chapter 17:	Feed Lines	181
Chapter 18:	Tests and Test Equipment	189
Chapter 19:	Interference and Grounding	193
Chapter 20:	Batteries and Other Power Sources	200
Chapter 21:	Connectors and Connecting Cables	207
Chapter 22:	RMS and PEP	210
Chapter 23:	Electrical and RF Safety	217

INTRODUCTION

This study guide is designed to teach you the material for the General Class FCC Amateur License exam. Each section of this book contains one portion of the material, and is followed by the actual questions that can appear on the test. The correct answers are shown in ***bold italic text***.

To earn you General license, you must pass the test for the Technician license, as well as this test. If you already have your Technician license, you will not need to take that test again. If you do not yet have your license, you can take both tests the same day, but you will need to take the Technician test first.

The Technician exam covers some of the same material, in a more elementary version. However, it also includes some material that is not covered on the General test. If you're starting from scratch, you should also study for the Technician exam.

The material in this study guide is presented in the most elementary fashion possible. In some cases, I'm probably guilty of over-simplification. However, by reading this book, you will be able to pass the test, and you confirm this by answering the practice questions at the end of each chapter. In many cases, reading the chapter a single time will be sufficient. However, a few chapters might require you to go back and review.

When you're finished, there might be one or two sections that you simply don't understand. If that's the case, don't worry. These are the exact questions that will be on the test. So if you need to simply memorize a few of the questions, that will get you through the test. There is one word of caution, however. If you memorize answers, you must memorize the complete answer, not just the letter for that answer. These same answers will be on the test, but they will not be in the same order.

The version of the test covered by this book is the 2019 version which is in effect until June 20, 2023. If you take the test after that date, these same questions will no longer be used. However, most questions generally stay the same, so even if you are using an old copy of this book, you will cover virtually all of the same material A new version of this guide will be published in 2023.

Plain-English Study Guide for the FCC Amateur Radio General Class License

1 IMPEDANCE AND REACTANCE

When you studied for your Technician license, you learned about Ohm's Law and the concept of resistance. **Resistance**, as you recall, is the opposition to the flow of electricity in a circuit, and is measured in Ohms.

There is a similar concept, which only applies to alternating current (AC) circuits, and that concept is **impedance**. Like resistance, impedance is measured in Ohms, and Ohm's law still applies. As you recall, Ohm's law is expressed in the following three formulas:

Volts equals Ohms times Amps

Ohms equals Volts divided by Amps

Amps equals Volts divided by Ohms.

Impedance consists of both resistance and **reactance**. All three are measured in Ohms. Resistance applies to both AC and DC circuits. **Reactance** applies only to AC circuits. Just like resistance, reactance is the the opposition to the flow of electricity. But reactance applies only to alternating current, and is caused by capacitance or inductance.

Impedance is the combination of both resistance and reactance, and is the total opposition to the flow of current in an AC circuit. (You cannot simply add up the resistance and reactance to get the total impedance.

Combining them uses a formula similar to the Pythagorean theorem, but you do not need to know that formula for the test.)

Because reactance is caused by inductance or capacitance, this means that every inductor and capacitor has a reactance that can be calculated or measured in ohms. This will vary depending on frequency, and will also

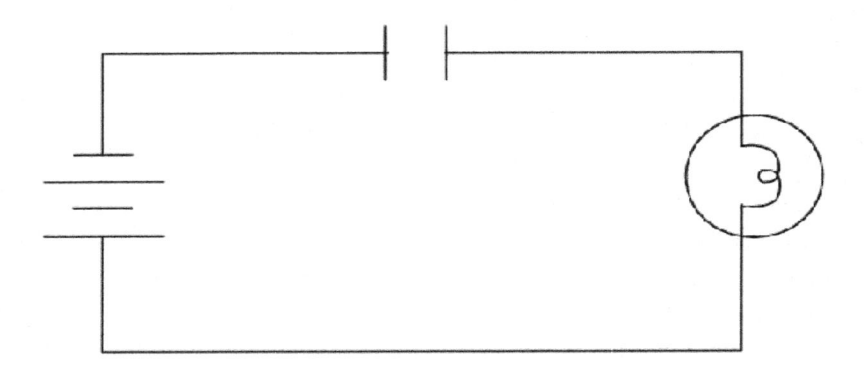

vary depending upon the inductance or capacitance.

Reactance is not difficult to understand. Think for a minute about what a capacitor is. As you probably remember from your Technician exam, capacitance is the ability to store energy in an electric field. A capacitor, in its simplest form, consists of two metal plates separated by air or some other insulator. Look at the following diagram, and it should be clear that the light bulb will not light:

The light will not come on, because there is no path for the DC electricity. To put it another way, the resistance of the capacitor is infinity.

Now, look at the following diagram, where an inductor has replaced the capacitor:

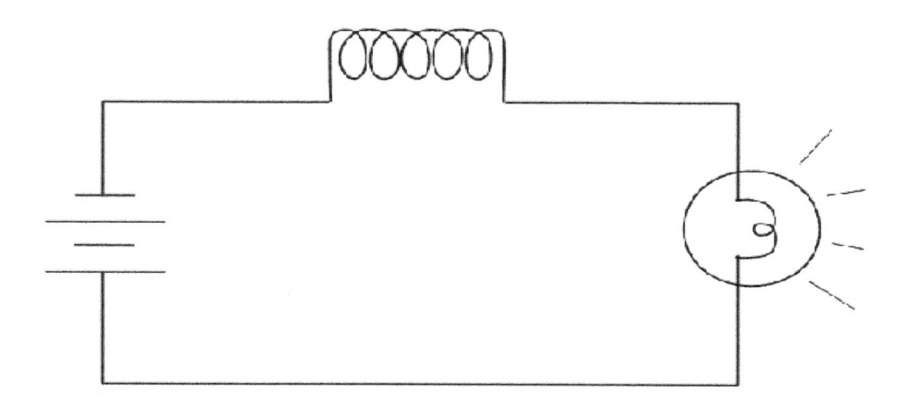

Plain-English Study Guide for the FCC Amateur Radio General Class License

As you recall, an inductor is nothing more than a coil of wire. From the DC electricity's point of view, the lamp is connected to the battery by nothing more than an ordinary piece of wire. The DC electricity doesn't care that the wire is coiled up. Wire is wire, and the inductor has a resistance of zero ohms.

Now, let's replace the flashlight bulb with a 120 volt bulb, and replace the battery with AC electricity. As you know, standard household current is Alternating Current, and its frequency is 60 Hz (60 cycles per second). This means that the current changes direction 120 times per second. Every second, it is positive for 1/60 second, and it's negative for 1/60 second. Here's what our light bulb and capacitor look like:

Richard P. Clem

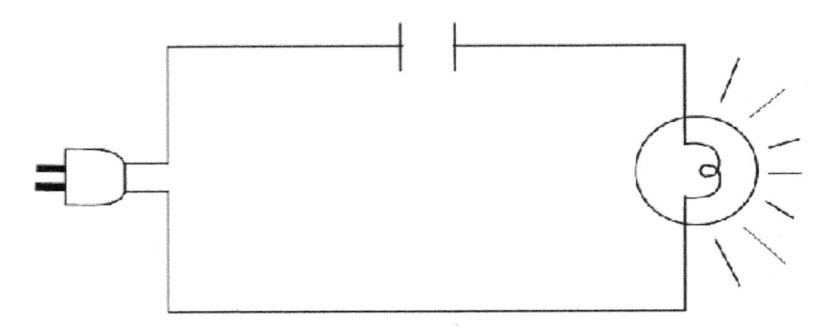

When you plug this in, there is no connection between the light bulb and the plug. Just like with the battery, the resistance of the capacitor is infinity. However, the electrons from the outlet want to get to the capacitor and charge up the plates. Remember, one side of the capacitor wants to charge up and become positive, and the other side wants to become negative. In order for this to happen, some of the electricity will flow through the bulb, and the bulb will light up for 1/120th of a second.

1/120th of a second later, the current from the outlet will change direction. At this point, the charge on the positive side of the capacitor will need to flow to the negative side of the outlet. Again, some of this current will need to flow through the bulb, and the bulb will light up for another 1/120th of a second.

This process keeps repeating itself, and every time the current changes direction, some of the current will flow through the bulb. Because the current is flowing through it, the bulb will light up. In other words, current is flowing through this circuit, even though there's a gap in the middle of the capacitor. The gap will block some of the flow, but it will not block all of it. In other words, it works just like resistance, and can be measured in ohms, just like a resistor. This "resistance" is called **capacitive reactance**.

Plain-English Study Guide for the FCC Amateur Radio General Class License

Take a look again at the first diagram. As you remember, the bulb will not light because the DC electricity cannot flow through the capacitor. In other words, with DC, the capacitor acts just like an insulator, or a resistor with a resistance of infinity ohms. You can think of DC as having a frequency of zero. Remember, frequency is the number of times per second that the electricity changes direction. In the case of DC, it changes directions zero times per second, so the frequency is zero Hertz.

So we know that when the frequency is zero, the capacitive reactance is infinity. But we also know that when the frequency is higher than zero, the reactance goes down. It never gets to zero, but it keeps going down as we increase the frequency. For the test, you do not need to know the formula for calculating this. But you do need to know that as frequency increases, capacitive reactance decreases.

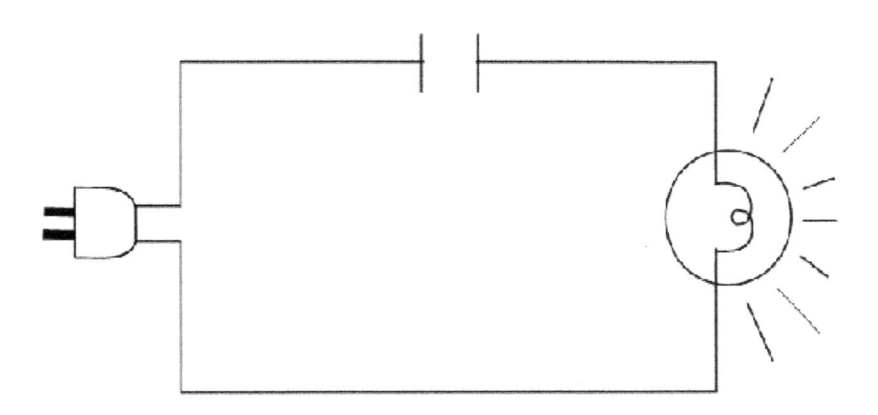

In the case of an inductor, the process is just the opposite. As you recall, from the second diagram, when working with DC (frequency zero), the inductor has absolutely no effect. It is just like a resistor of zero ohms. But if we hook up a similar circuit using AC, the inductor will begin to have an effect. Take a look at the following diagram:

Richard P. Clem

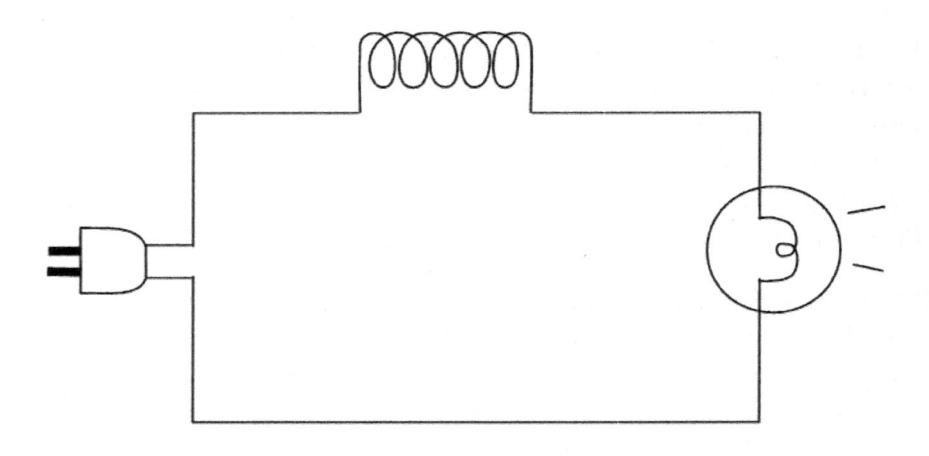

When the power is first hooked up, some of the energy is used to form a magnetic field around the coil. If you hold a compass near the coil, you would be able to see this field. Some of the energy is used to form this field.

But 1/120th second later, the current changes direction. As a result, the magnetic field changes polarity. Some of the electrical energy is used to take down the old magnetic field and put up the new field in the opposite direction. The energy to do this must come from somewhere, and as a result, the bulb will burn more dimly than it would otherwise. In other words, the inductor is acting just like a resistor, and this effect can be measured in Ohms. At a frequency of zero Hertz (DC), there is no effect, and this is zero ohms. As the frequency increases, the number of ohms increases. This property (the number of Ohms) is called **inductive reactance**. As the frequency increases, the inductive reactance increases.

You do not need to know the exact formulas for the test, but for the test, you do need to know the following facts about reactance:

Plain-English Study Guide for the FCC Amateur Radio General Class License

In an inductor, as the frequency goes up, the reactance also goes up. (You also need to know for the test that if the frequency goes too high, past the inductor's "self-resonant frequency," then it will start to act like a capacitor.)

In a capacitor, as the frequency goes up, the reactance goes down.

If you forget these two facts, just think about what happens when the frequency is zero. In the capacitor, the number of Ohms is infinity, because a capacitor is just two metal plates that are not connected. In an inductor, the number of Ohms is zero, because an inductor is just a piece of wire, and the DC electricity does not care that the wire is coiled up.

There are questions on the test about **impedance matching**, but the questions do not go into detail. The main thing that you need to know is that when the impedance of a source and a load are matched (in other words, the same number of ohms), then there will be the maximum transfer of power.

One way to match the impedance is to insert an LC network between the source and the load. An "LC network" is simply an inductor (coil) and a capacitor. This might also be called a "Pi network", because the coil and capacitors might be arranged to look like the Greek letter Pi.

Another method would be an impedance matching transformer.

If you are matching an antenna to a transmitter, one method of impedance matching might even include using a transmission line of a particular length.

What is impedance?

A. The electric charge stored by a capacitor

B. The inverse of resistance

C. The opposition to the flow of current in an AC circuit

D. The force of repulsion between two similar electric fields

What is reactance?

A. Opposition to the flow of direct current caused by resistance

B. Opposition to the flow of alternating current caused by capacitance or inductance

C. A property of ideal resistors in AC circuits

D. A large spark produced at switch contacts when an inductor is de-energized

Which of the following causes opposition to the flow of alternating current in an inductor?

A. Conductance

B. Reluctance

C. Admittance

D. Reactance

Which of the following causes opposition to the flow of alternating current in a capacitor?

A. Conductance

B. Reluctance

C. Reactance

D. Admittance

Plain-English Study Guide for the FCC Amateur Radio General Class License

How does an inductor react to AC?

A. As the frequency of the applied AC increases, the reactance decreases

B. As the amplitude of the applied AC increases, the reactance increases

C. As the amplitude of the applied AC increases, the reactance decreases

D. As the frequency of the applied AC increases, the reactance increases

How does a capacitor react to AC?

A. As the frequency of the applied AC increases, the reactance decreases

B. As the frequency of the applied AC increases, the reactance increases

C. As the amplitude of the applied AC increases, the reactance increases

D. As the amplitude of the applied AC increases, the reactance decreases

What unit is used to measure reactance?

A. Farad

B. Ohm

C. Ampere

D. Siemens

What happens when the impedance of an electrical load is equal to the internal impedance of the power source, assuming both impedances are resistive?

A. The source delivers minimum power to the load

B. The electrical load is shorted

C. No current can flow through the circuit

D. *The source can deliver maximum power to the load*

Which of the following describes one method of impedance matching between two AC circuits?

A. *Insert an LC network between the two circuits*

B. Reduce the power output of the first circuit

C. Increase the power output of the first circuit

D. Insert a circulator between the two circuits

What is one reason to use an impedance matching transformer?

A. To minimize transmitter power output

B. *To maximize the transfer of power*

C. To reduce power supply ripple

D. To minimize radiation resistance

Which of the following devices can be used for impedance matching at radio frequencies?

A. A transformer

B. A Pi-network

C. A length of transmission line

D. All of these choices are correct

What happens when an inductor is operated above its self-resonant frequency?

A. Its reactance increases

B. Harmonics are generated

C. It becomes capacitive

D. Catastrophic failure is likely

Richard P. Clem

2 TRANSFORMERS

The test covers a number of questions about various electronic components, and the first of these is the transformer.

A **transformer** is simply two coils of wire placed next to each other. When AC power is connected to one of the coils, this causes a magnetic field. The magnetic field can cause an electric current to flow in the other coil. This phenomenon is called **mutual inductance**. The coil that is actually hooked up to the power source is called the **primary**. The other coil is called the **secondary**.

Here is the schematic symbol for a transformer:

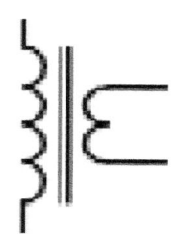

Plain-English Study Guide for the FCC Amateur Radio General Class License

TRANSFORMER

One common use of a transformer is to change voltage. When one AC voltage is hooked up to the primary, another voltage will appear on the secondary. This depends on the ratio of the number of winding turns. For example, if there are 100 turns on the primary, and 1000 turns in the secondary, then the voltage will go up by a factor of ten: If the primary is hooked up to 120 volts, then the secondary voltage would be 1200 volts.

If the primary was 1000 turns and the secondary was 100 turns, then the voltage would go down by a factor of 10. In other words, if the primary is hooked up to 120 volts, then the secondary voltage would be 12 volts.

To put this another way, the ratio of voltages is exactly the same as the ratio of turns. If the secondary of the transformer is twice as big as the primary, then the voltage will be twice as big. If the secondary has three times the number of turns as the primary, then the secondary voltage will be three times as high.

If you reverse a transformer, then it becomes the opposite of what it was before. In other words, if the transformer was originally a "step down" transformer that dropped the voltage by one fourth, then it will become a "step up" transformer and increase the voltage four times.

If the voltage goes down in one side of a transformer, that means that the current in that side needs to go up. Therefore, it's common for the side of the transformer with fewer windings (the low-voltage side) to have thicker wire, because it needs to handle more current.

The questions on the test are relatively simple, and can be figured out if you read them carefully and remember that the ratio of voltage is the same as the ratio of turns. It's probably not very helpful to memorize a formula. If you remember what is going on, you can easily figure out these questions.

As we learned earlier, a transformer can also be used to match impedance. The math is only slightly more complicated, but still not very difficult. In the case of impedance, you need to remember that the turns ratio is the square root of the impedance ratio. Or to put it another way, the turns ratio multiplied by itself equals the impedance ratio.

For example, if the primary has 100 turns, and the secondary has 500, this is a turns ratio of 5. Therefore, the impedance would change by a factor 25, because 5 x 5 = 25.

There is only one question on the test about using a transformer to match impedance. The question states that you need a transformer to match a 4 ohm speaker with a 600 ohm amplifier. So the impedance ratio is 150 (600 divided by 4). The turns ratio you need is the square root of 150, which is about 12.2. (12.2 times 12.2 equals approximately 150).

What causes a voltage to appear across the secondary winding of a transformer when an AC voltage source is connected across its primary winding?

A. Capacitive coupling

B. Displacement current coupling

C. Mutual inductance

D. Mutual capacitance

What happens if a signal is applied to the secondary winding of a 4:1 voltage step-down transformer instead of the primary winding?

A. The output voltage is multiplied by 4

B. The output voltage is divided by 4

C. Additional resistance must be added in series with the primary to prevent overload

D. Additional resistance must be added in parallel with the secondary to prevent overload

What is the RMS voltage across a 500-turn secondary winding in a transformer if the 2250-turn primary is connected to 120 VAC?

A. 2370 volts

B. 540 volts

C. 26.7 volts

D. 5.9 volts

What is the turns ratio of a transformer used to match an audio amplifier having a 600 ohm output impedance to a speaker having a 4 ohm impedance?

A. 12.2 to 1

B. 24.4 to 1

C. 150 to 1

D. 300 to 1

Why is the conductor of the primary winding of many voltage step up transformers larger in diameter than the conductor of the secondary winding?

A. To improve the coupling between the primary and secondary

B. To accommodate the higher current of the primary

C. To prevent parasitic oscillations due to resistive losses in the primary

D. To ensure that the volume of the primary winding is equal to the volume of the secondary winding

Plain-English Study Guide for the FCC Amateur Radio General Class License

(circle diagram with V on top, I and R on bottom)

3 CALCULATING POWER

For the Technician exam, you learned how to calculate power. As you probably remember, power is calculated from the following formula:

Watts equals volts times amps.

This formula is not covered directly on the General exam, but there will be some slightly more complicated questions about power. For the Technician exam, the question gave you the number of volts and amps, and you simply had to multiply them together and get the number of watts.

For the General exam, you will need to calculate power, but you will not be given both volts and amps. One of these numbers will be missing, but the question will include a number of Ohms.

The easiest way to approach these problems is to use two steps. First, find the missing item, which will be either volts or amps. After you have found this, then just multiply volts times amps to get the final answer.

The first step (finding the missing element) will require you to use Ohm's law. To review from the Technician exam, here is Ohm's Law:

Volts equals Ohms times Amps. $V = IR$

Ohms equals Volts divided by Amps. $R = V/I$

$I = V/R$

Richard P. Clem

$I = V/R$

Amps equals Volts divided by Ohms.

So if the problem is asking you to calculate power but only gives you Volts and Ohms, your first step is to find the number of Amps. If the problem tells you Amps and Ohms, then your first step is to find the number of Volts. After you have done this, multiply Volts times Amps to get Watts.

To see how to do this, let's work through two of the problems from the test:

How many watts of electrical power are used if 400 VDC is supplied to an 800-ohm load?

We know the voltage, 400, but we don't know the number of amps. So first, we need to calculate the number of amps. From Ohm's Law, we know that Amps equals Volts divided by Ohms. Here, that is 400 divided by 800, so the current is 0.5 Amps. Now that we know this, we can easily calculate the power: Watts equals Volts times Amps. 400 times 0.5 equals 200, so the correct answer is 200 Watts.

Here's another question right from the test: How many watts are dissipated when a current of 7.0 milliamperes flows through 1.25 kilohms? Here, the first step is to find the voltage. From Ohm's law, we know that Volts equals Amps times Ohms. Watch out here, because we're dealing with milliamps (1/1000 Amp) and kilohms (1000 Ohms). So to calculate the voltage, simply enter the following into your calculator: 7 divided by 1000 times 1.25 times 1000, which equals 8.75 volts.

Finally, to get power, we use the formula watts equals volts times amps. Remember that we're still dealing with milliamps, which are 1/1000 amp. So to calculate the power, it is 8.75 times 7 divided by 1000, which is 0.06125. So the correct answer is 0.06125 Watts. We look at the possible answers, and one of them is "approximately 61 milliwatts". Since a milliwatt is 1/1000 watt, we know that 61 milliwatts is the same as 61 divided by 1000 watts. We punch this into the calculator, and it shows 0.061, which is approximately the same as our answer. So we know this is the right answer.

$P = VI = V\left(\dfrac{V}{R}\right) = \dfrac{V^2}{R} = (IR)I = I^2 R$

Plain-English Study Guide for the FCC Amateur Radio General Class License

This method will always work. Since you already know both of these formulas from the Technician exam, this is probably the best way to do these problems. However, if you want to, you can save some work by memorizing the following formulas:

Power equals voltage squared divided by resistance. $P = V^2/R$

Power equals current squared times resistance. $P = I^2 R$

But you do not need to memorize these formulas for the test. The two-step method described above will always work.

There is one more question on the test about basic electricity. You need to know that in a circuit with more than one parallel branch, then the total current is equal to the sum of the currents in each branch. (This principle is known as Kirchhoff's current law, but you do not need to know that name for the test.)

How many watts of electrical power are used if 400 VDC is supplied to an 800-ohm load?

A. 0.5 watts

B. 200 watts

C. 400 watts

D. 3200 watts

$P = \dfrac{V^2}{R} = \dfrac{(400)^2}{800} = \dfrac{400(400)}{800} = \dfrac{400}{2} = 200$

How many watts of electrical power are used by a 12-VDC light bulb that draws 0.2 amperes?

$P = VI = 12(0.2) = 2.4$

A. 2.4 watts

B. 24 watts

C. 6 watts

D. 60 watts

How many watts are dissipated when a current of 7.0 milliamperes flows through a 1250 ohm resistance?

A. Approximately 61 milliwatts

B. Approximately 61 watts

C. Approximately 11 milliwatts

D. Approximately 11 watts

$P = i^2 R = (0.007)^2 (1250)$

$P = 0.061125 \text{ W}$

$P = 61.25 \text{ mW}$

$P \approx 61 \text{ mW}$

How does the total current relate to the individual currents in each branch of a purely resistive parallel circuit?

A. It equals the average of each branch current

B. It decreases as more parallel branches are added to the circuit

C. It equals the sum of the currents through each branch

D. It is the sum of the reciprocal of each individual voltage drop

Plain-English Study Guide for the FCC Amateur Radio General Class License

4 COMBINING COMPONENTS

It frequently comes up as a Ham that you are working on something, need a part, such as a resistor, capacitor, or inductor, but don't have the exact value you need. You can usually get the value you need by combining one or more components. For example, if you need a 200 ohm resistor but don't have one, you can combine two 100 ohm resistors in series:

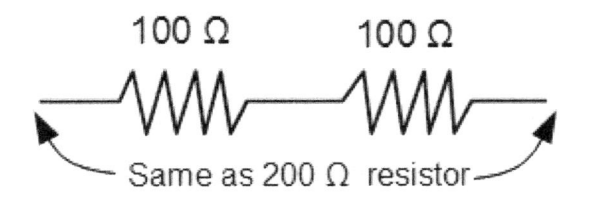

100 Ω   100 Ω
—/\/\/\—/\/\/\—
Same as 200 Ω resistor

Inductors work exactly the same way: They add up when connected in series.

Capacitors work differently. They add up when connected in **parallel**:

50 µF

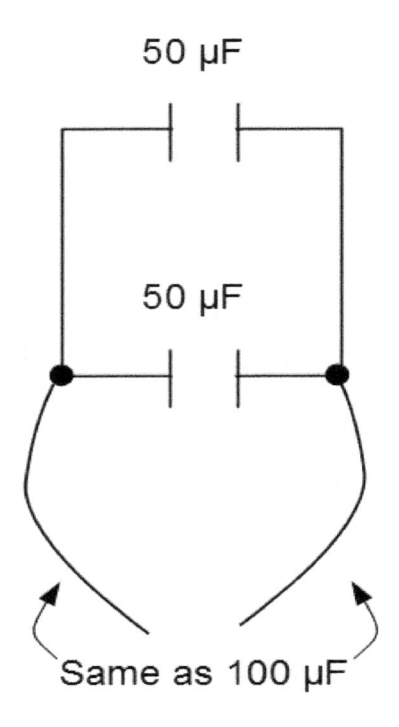

50 µF

Same as 100 µF

The math is slightly more complicated dealing with resistors in parallel. The first thing to remember is that the resistance always goes down: **The combined resistance of two or more resistors in parallel is always less than the smallest resistor.** If you remember this fact, you will be able to answer most of these questions. This makes sense, because the electricity has an additional path it can follow. Here are two resistors in parallel:

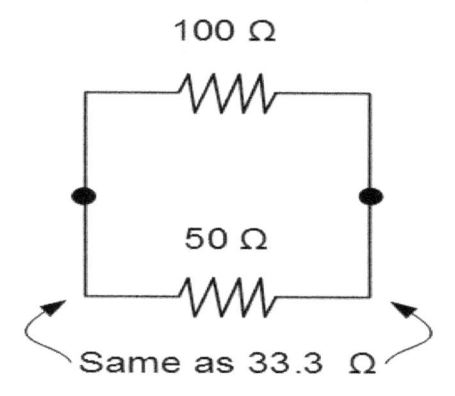

Plain-English Study Guide for the FCC Amateur Radio General Class License

```
        100 Ω
        ─WWW─
    ┌───       ───┐
    •              •
    │    50 Ω     │
    └───  WWW  ───┘
      Same as 33.3 Ω
```

One resistor is 100 ohms, and the other one is 50 ohms. The combined resistance is less than the smallest resistor, and works out to 33.3 ohms. To calculate the combined resistance of **two** resistors in parallel, you use the "product over the sum" formula. You multiply the two values and get the product, in this case, get 100 x 50 = 5000. The sum is the two numbers added together, 100 + 50 = 150. The final answer is the product divided by the sum: 5000 divided by 150, which equals about 33.3. So the answer is 33.3 Ohms.

This method only works with two resistors. If you have three or more resistors, you have to work on them two at a time. You can **not** multiply all three, and divide that by the sum or all three. If you have a question with more than one resistor in parallel, divide them into groups of two. Figure out the combined resistance of those two, and then combine it with the other one, using the same process.

Remember, inductors follow the same rules as resistors when it comes to combining them. So two inductors in parallel follow the "product over the sum" rule.

Capacitors in **series** also follow the "product over the sum" rule: If there are two capacitors in series, the combined capacitance will be less than any of the original capacitors.

For the technician exam, you had to learn many metric prefixes, such as kilo-, micro-, etc. For the general exam, you need to know one more, the prefix **nano-**, which means one billionth. You need to be able to change from picoFarads (pF) to nanoFarads (nF). To go from pico to nano, divide by 1000. To go from nano to pico, multiply by 1000.

Which of the following components increases the total resistance of a resistor?

A. A parallel resistor

B. A series resistor

C. A series capacitor

D. A parallel capacitor

What is the total resistance of three 100-ohm resistors in parallel?

A. .30 ohms

B. .33 ohms

C. 33.3 ohms

D. 300 ohms

If three equal value resistors in series produce 450 ohms, what is the value of each resistor?

A. 1500 ohms

B. 90 ohms

C. 150 ohms

D. 175 ohms

What is the value in nanofarads (nF) of a 22,000 pF capacitor?
A. 0.22 nF
B. 2.2 nF
C. 22 nF
D. 220 nF

What is the value in microfarads of a 4700 nanofarad (nF) capacitor?
A. 47 µF
B. 0.47 µF
C. 47,000 µF
D. 4.7 µF

What is the equivalent capacitance of two 5.0 nanofarad capacitors and one 750 picofarad capacitor connected in parallel?

A. 576.9 nanofarads

B. 1733 picofarads

C. 3583 picofarads

D. 10.750 nanofarads

Richard P. Clem

What is the capacitance of three 100 microfarad capacitors connected in series?

A. .30 microfarads

B. .33 microfarads

C. 33.3 microfarads

D. 300 microfarads

What is the inductance of three 10 millihenry inductors connected in parallel?

A. .30 Henrys

B. 3.3 Henrys

C. 3.3 millihenrys

D. 30 millihenrys

What is the inductance of a 20 millihenry inductor connected in series with a 50 millihenry inductor?

A. 0.07 millihenries

B. 14.3 millihenries

C. 70 millihenries

D. 1000 millihenries

Plain-English Study Guide for the FCC Amateur Radio General Class License

What is the capacitance of a 20 microfarad capacitor in series with a 50 microfarad capacitor?

A. .07 microfarads

B. 14.3 microfarads

C. 70 microfarads

D. 1000 microfarads

Which of the following components should be added to a capacitor to increase the capacitance?

A. An inductor in series

B. A resistor in series

C. A capacitor in parallel

D. A capacitor in series

Which of the following components should be added to an inductor to increase the inductance?

A. A capacitor in series

B. A resistor in parallel

C. An inductor in parallel

D. An inductor in series

What is the total resistance of a 10 ohm, a 20 ohm, and a 50 ohm resistor in parallel?

A. 5.9 ohms

B. 0.17 ohms

C. 10000 ohms

D. 80 ohms

Plain-English Study Guide for the FCC Amateur Radio General Class License

5 DECIBELS

One unit that you need to be aware of for the test is the **decibel (dB)**. A decibel is a unit used to measure the ratio of two different quantities, usually power. The main fact that you need to know is that anytime the power doubles, this is an increase of 3 dB. So if the power goes from one watt to two watts, this is an increase of 3 dB. If it doubles twice (in other words, it goes from the original value to four times as much), this would be a total of 6 dB.

For example, if we start with 10 watts, and we want to know how many dB increase it would be if it went up to 40 watts, here is how you would do the problem. First, you increase it to 20 watts. When this happens, it went up 3 dB, because it doubled. (Remember, that's the one fact that you should memorize—doubling is the same as a 3 dB increase). Then, you increase it again to 40 watts. This increased it by 3 more dB, for a total of 6 dB.

If it went up by a factor of 8 (for example, it went from 100 watts to 800 watts), this would be about 9 dB, because we had to double it three times (first from 100 to 200, then from 200 to 400, and then from 400 to 800).

You should be able to solve all but one of the problems this way. But another fact that will make some of the problems easier is that an increase by a factor of 10 is equal to exactly 10 dB. For example, if we went from

100 watts to 1000 watts, this would be 10 dB. (Note that this is similar to the last problem—an increase from 100 watts to 800 watts is about 9 dB, so it makes sense that this answer will be slightly higher, which it is.)

One item that is usually measured in dB is the amount of loss of a particular kind of feed line. It is usually measured in dB per hundred feet. So if you have a 50 foot piece of cable with a loss of 6 dB per hundred feet, then the loss would be about 3 dB. In other words, the signal at the output would be about half as strong as the input.

Decibels are mentioned in the questions about S-meters. An S-meter is the meter on a receiver that measures received signal strength. Generally, these meters run from zero to nine, and above "S-9", they are calibrated in decibels. The question on the test refers to "20 dB above S-9", and asks how much stronger this is than S-9.

There are two ways you can do this problem. 20 is approximately 21, and 21 = 3+3+3+3+3+3+3. Therefore, we know that the signal doubled seven times. In other words, it is 2 x 2 x 2 x 2 x 2 x 2 x 2 = 128 times the original signal. The closest answer on the test is 100, so this is the right answer.

A slightly easier way to do this problem (and one that will give the exact answer, not an approximation) is as follows: 10 dB is the same as an increase by a factor of 10. In the problem on the test, the signal went up by 20 dB, which means it went up by 10 dB once, and then went up by 10 dB a second time. The first time, it got multiplied by 10. And the second time, it got multiplied by 10 again. So the final signal is 10 x 10 = 100 times stronger than the original.

On most receivers, each "S-unit" is about 6 dB, meaning that each S-unit represents a signal four times stronger than the S-unit below it. So a signal of S-9 is about four times stronger than a signal of S-8.

There is one question about decibels that is difficult to answer using the simplified method described above. You can simply memorize that answer: **A 1 dB reduction in power equals a reduction of 20.5%.**

Plain-English Study Guide for the FCC Amateur Radio General Class License

Finally, if you plan to use a calculator on the test and want to memorize the formula that can be used for any of the decibel problems, here is the formula.

$$dB = 10 \log (P1/P2)$$

P1 and P2 are the two values that you are comparing, and the log is the base 10 logarithm on your calculator. But if you don't want to use this formula, remember, you can use two rules to solve all but one of the problems on the test: First, 3 dB is the same as doubling, and 10 dB is the same as increasing by a factor of 10.

Some of the antenna questions refer to "dBi" and "dBd." All you need to know for the test is that dBi means the gain of antenna compared to an isotropic radiator, dBd means the gain compared to a dipole, and dBi is always 2.5 dB higher than dBd.

What dB change represents a factor of two increase or decrease in power?

A. Approximately 2 dB

B. Approximately 3 dB

C. Approximately 6 dB

D. Approximately 12 dB

In what units is RF feed line loss usually expressed?

A. Ohms per 1000 feet

B. Decibels per 1000 feet

C. Ohms per 100 feet

D. Decibels per 100 feet

What percentage of power loss would result from a transmission line loss of 1 dB?

A. 10.9 percent

B. 12.2 percent

C. 20.6 percent

D. 25.9 percent

What does an S meter measure?

A. Conductance

B. Impedance

C. Received signal strength

D. Transmitter power output

How does a signal that reads 20 dB over S9 compare to one that reads S9 on a receiver, assuming a properly calibrated S meter?

A. It is 10 times less powerful

B. It is 20 times less powerful

C. It is 20 times more powerful

D. It is 100 times more powerful

Plain-English Study Guide for the FCC Amateur Radio General Class License

Where is an S meter found?

A. In a receiver

B. In an SWR bridge

C. In a transmitter

D. In a conductance bridge

How much must the power output of a transmitter be raised to change the S- meter reading on a distant receiver from S8 to S9?

A. Approximately 1.5 times

B. Approximately 2 times

C. Approximately 4 times

D. Approximately 8 times

How does antenna gain stated in dBi compare to gain stated in dBd for the same antenna?

A. dBi gain figures are 2.15 dB lower then dBd gain figures

B. dBi gain figures are 2.15 dB higher than dBd gain figures

C. dBi gain figures are the same as the square root of dBd gain figures multiplied by 2.15

D. dBi gain figures are the reciprocal of dBd gain figures + 2.15 dB

What is meant by the terms dBi and dBd when referring to antenna gain?

A. dBi refers to an isotropic antenna, dBd refers to a dipole antenna

B. dBi refers to an ionospheric reflecting antenna, dBd refers to a dissipative antenna

C. dBi refers to an inverted-vee antenna, dBd refers to a downward reflecting antenna

D. dBi refers to an isometric antenna, dBd refers to a discone antenna

6 OTHER COMPONENTS

The test covers a few miscellaneous facts about various components.

There are a number of different kinds of capactitors used for different purposes. One of the lowest cost capacitors is the **certamic disc** capacitor. The type of capacitor with the highest capacitance for its size is generally the **electrolytic** capacitor. Electrolytic capacitors are often used to filter the output of a power supply. In this application, it's important that the capacitor has a low equivalent series resistance. With electrolytic capacitors, it's very important to hook them up with the correct polarity. If the polarity is wrong, this can cause a short circuit and destroy the dialectric layer. It can even cause the capacitor to overheat and explode.

When working with capacitors in VHF and UHF circuits, you need to be careful to keep lead lengths short. This is because the leads my add inductance, which would reduce the effective capacitance.

Resistors are frequently made of carbon. Often, the resistance will change based on temperature. The amount of change will depend on the resistor's **temperature coefficient**. A **thermistor** is a special kind of resistor that has very specific changes in resistance due to temperature. It would be used for measuring temperature.

Richard P. Clem

Some resistors are **wire-wound**. These are typically a bad choice for RF (Radio Frequency) circuits, because the windings would add inductance to the circuit, which might be unpredictable.

Inductors are often wound on **torroidal** ferrite forms. "Torroid" simply means donut shaped. **Ferrite** is a material with good magnetic properties for winding coils. The performance of a ferrite core at different frequencies depends on the composition or "mix" of the materials used. You also need to know that ferrite beads or cores can reduce RF current on the shield of coaxial cable. It does this by creating an impedance.

To minimize mutual inductance between two inductors, they should be at right angles. Mutual inductance can be unwanted many places, because it might cause an unwanted interaction between two stages of a circuit.

An inductor is sometimes used to smooth the output of a DC power supply. When an inductor is used this way, it is often called a **filter choke**.

There is one question about inter-turn capacitance of an inductor. The correct answer points out that this might cause **self-reasonance** at some frequencies.

Which of the following is an advantage of ceramic capacitors as compared to other types of capacitors?

A. Tight tolerance

B. High stability

C. High capacitance for given volume

D. *Comparatively low cost*

Which of the following is an advantage of an electrolytic capacitor?

A. Tight tolerance

B. Non-polarized

C. High capacitance for given volume

D. Inexpensive RF capacitor

Why is the polarity of applied voltages important for polarized capacitors?

A. Incorrect polarity can cause the capacitor to short-circuit

B. Reverse voltages can destroy the dielectric layer of an electrolytic capacitor

C. The capacitor could overheat and explode

D. All of these choices are correct

Which of the following is a reason not to use wire-wound resistors in an RF circuit?

A. The resistor's tolerance value would not be adequate for such a circuit

B. The resistor's inductance could make circuit performance unpredictable

C. The resistor could overheat

D. The resistor's internal capacitance would detune the circuit

What is an advantage of using a ferrite core toroidal inductor?

A. Large values of inductance may be obtained

B. The magnetic properties of the core may be optimized for a specific range of frequencies

C. Most of the magnetic field is contained in the core

D. All of these choices are correct

What determines the performance of a ferrite core at different frequencies?

A. Its conductivity

B. Its thickness

C. The composition, or "mix," of materials used

D. The ratio of outer diameter to inner diameter

How does a ferrite bead or core reduce common-mode RF current on the shield of a coaxial cable?

A. By creating an impedance in the current's path

B. It converts common-mode current to differential mode

C. By creating an out-of-phase current to cancel the common-mode current

D. Ferrites expel magnetic fields

Plain-English Study Guide for the FCC Amateur Radio General Class License

As you remember from your Technician exam, a **rectifier** is a device that allows current to flow in only one direction, and is used to change AC to DC. When you buy a rectifier, one of the specifications it will have is the **peak inverse voltage**. This means the maximum voltage it will handle in the non-conducting direction.

Another rating that a rectifier will have is the **average forward current**.

A semiconductor diode won't "kick in" until the voltage meets a certain threshold. In the case of a germanium diode, this value is about 0.3 volts. In a silicon diode, this value is about 0.7 volts.

Sometimes, it is necessary to hook two diodes in parallel to increase the current handling capacity. When you do this, a resistor is connected in series with each diolde to ensure that one diode doesn't carry most of the current.

A **Schottky diode** is sometimes used in an RF switching circuit because it has a lower capacitance than a silicon diode.

One special type of diode is the **LED (Light Emitting Diode)**. As the name implies, this type of diode gives off light when energized. Specifically, it emits light when **forward biased** (in other words, when current is passing through it). LED's have replaced incandescent light bulbs in many applications because incandescent bulbs have the disadvantage of high power consumption.

(Many radios have gone a step further and use liquid-crystal displays. However, these displays require a separate light source, in the form of ambient or back lighting.)

Transistors can be used for many purposes. They can be used as amplifiers and oscillators. They can be used for switching in a logic circuit. In a logic circuit, the stable operating points are the saturation and cut-off regions.

Many large power transistors use the outer metal case of the transistor as one of the three leads. The metal case is usually the collector (of a bipolar junction transistor) or the drain (of an FET). When you mount the transistor, it must be insulated from ground to keep shorting these voltages.

A **MOSFET** is a transistor whose gate is separated from the channel with a thin insulating layer.

A **vacuum tube** consists of a **cathode, grid**, and **plate**. Electrons flow from the cathode, through the grid, and to the plate. Some tubes have more than one grid. In this case, one of them is called the **control grid**, and this is the grid used to control the flow of electrons. (Just remember that "control" and "regulate" mean the same thing, and the correct answer to this question will be obvious.) Another grid, the **screen grid**, is designed to reduce grid-to-plate capacitance.

A vacuum tube operates almost identically to a Field Effect Transistor (FET).

You need to know the following additional schematic symbols, in addition to the ones learned earlier:

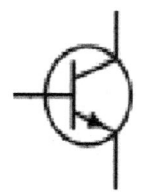

| FET (FIELD EFFECT TRANSISTOR | NPN TRANSISTOR | ZENER DIODE | TAPPED INDUCTOR |

Plain-English Study Guide for the FCC Amateur Radio General Class License

There are a few questions on the test about digital components. A **microprocessor** is a computer on a single integrated circuit. A **microcontroller** can often replace complex digital circuitry. **Non-volatile** memory means that the stored information is maintained even if the power is removed.

There are two types of digital integrated circuits referred to on the test: **CMOS** and **TTL**. CMOS has the advantage of low power consumption.

The test has one question each about **AND** gates and **NOR** gates. In an AND gate, the output is high only when both inputs are high. In a NOR gate, the output is low when either or both inputs are high.

There are two types of analog integrated circuits referred to on the test: A **linear voltage regulator**, and an **operational amplifier**. You need to be able to identify these as being analog.

$$AND \begin{cases} 0.0 = 0 \\ 0.1 = 0 \\ 1.0 = 0 \\ 1.1 = 1 \end{cases} \quad OR \begin{cases} 0+0 = 0 = 1 \\ 0+1 = 1 = 0 \\ 1+0 = 1 = 0 \\ 1+1 = 1 = 0 \end{cases} \begin{matrix} \overline{OR} \\ = \\ NOR \end{matrix}$$

What is the approximate junction threshold voltage of a germanium diode?

A. 0.1 volt

B. 0.3 volts   Ge → 0.3 V

C. 0.7 volts

D. 1.0 volts

What is the approximate junction threshold voltage of a conventional silicon diode?

A. 0.1 volt   Si → 0.7 V

B. 0.3 volts

C. 0.7 volts

41

D. 1.0 volts

What are the stable operating points for a bipolar transistor used as a switch in a logic circuit?

A. Its saturation and cutoff regions

B. Its active region (between the cutoff and saturation regions)

C. Its peak and valley current points

D. Its enhancement and deletion modes

Which of the following describes the construction of a MOSFET?

A. The gate is formed by a back-biased junction

B. The gate is separated from the channel with a thin insulating layer

C. The source is separated from the drain by a thin insulating layer

D. The source is formed by depositing metal on silicon

Which element of a triode vacuum tube is used to regulate the flow of electrons between cathode and plate?

A. Control grid

B. Heater

C. Screen Grid

D. Trigger electrode

Plain-English Study Guide for the FCC Amateur Radio General Class License

What is the primary purpose of a screen grid in a vacuum tube?

A. To reduce grid-to-plate capacitance

B. To increase efficiency

C. To increase the control grid resistance

D. To decrease plate resistance

How is an LED biased when emitting light?

A. Beyond cutoff

B. At the Zener voltage

C. Reverse Biased

D. Forward Biased

Which of the following is a characteristic of a liquid crystal display?

A. It utilizes ambient or back lighting

B. It offers a wide dynamic range

C. It consumes relatively high power

D. It has relatively short lifetime

Richard P. Clem

The following questions refer to the following diagram:

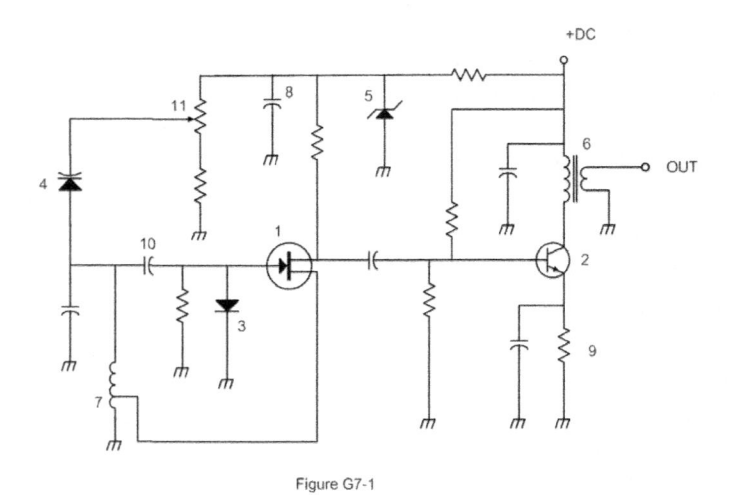

Figure G7-1

Which symbol in figure G7-1 represents a field effect transistor?

A. Symbol 2

B. Symbol 5

C. Symbol 1

D. Symbol 4

Which symbol in figure G7-1 represents a Zener diode?

Plain-English Study Guide for the FCC Amateur Radio General Class License

A. Symbol 4

B. Symbol 1

C. Symbol 11

D. Symbol 5

Which symbol in figure G7-1 represents an NPN junction transistor?

A. Symbol 1

B. Symbol 2

C. Symbol 7

D. Symbol 11

Which symbol in Figure G7-1 represents a multiple-winding transformer?

A. Symbol 4

B. Symbol 7

C. Symbol 6

D. Symbol 1

Which symbol in Figure G7-1 represents a tapped inductor?

A. Symbol 7

B. Symbol 11

C. Symbol 6

D. Symbol 1

Which of the following is an advantage of CMOS integrated circuits compared to TTL integrated circuits?

A. Low power consumption

B. High power handling capability

C. Better suited for RF amplification

D. Better suited for power supply regulation

What is meant when memory is characterized as "non-volatile"?

A. It is resistant to radiation damage

B. It is resistant to high temperatures

C. The stored information is maintained even if power is removed

D. The stored information cannot be changed once written

What kind of device is an integrated circuit operational amplifier?

A. Digital

B. MMIC

C. Programmable Logic

D. Analog

Which of the following describes the function of a two input AND gate?

A. Output is high when either or both inputs are low

B. Output is high only when both inputs are high

C. Output is low when either or both inputs are high

D. Output is low only when both inputs are high

Which of the following describes the function of a two input NOR gate?

A. Output is high when either or both inputs are low

B. Output is high only when both inputs are high

C. Output is low when either or both inputs are high

D. Output is low only when both inputs are high

Richard P. Clem

7 RADIO PROPAGATION

As you probably know, radio propagation is quite dependent upon the sun's activity. A key measure of this, that many Hams watch religiously, is the **sunspot number**. This is a measure of the solar activity based on counting sunspots and sunspot groups. High susnpot numbers enhance communications on the upper HF and lower VHF frequencies.

Three other related concepts are covered on the test: The **solar-flux index** is a measure of solar radiation measured at 10.7 cm. The **K-index** indicates the short term stability of the Earth's magnetic field. The **A-index** indicates the long term stability of the Earth's geomagnetic field.

The sun rotates on its axis once every 28 days. Therefore, HF propagation conditions often vary periodically in a 28-day cycle. The sun also has a cycle of approximately eleven years. At the peak of the eleven year sunspot cycle, sunspots are much more common than during the solar minimum. During periods of low solar activity, long distance communications on amateur frequencies 21 MHz and higher become much less reliable. The 20 meter band (14 MHz) usually supports daytime propagation at any point during the solar cycle.

A **sudden ionospheric disturbance** generally disrupts signals on lower frequencies more than on higher frequencies. A **geomagnetic storm** is a

Plain-English Study Guide for the FCC Amateur Radio General Class License

temporary disturbance in the Earth's ionosphere. One of these storms can degrade HF propagation, particularly at higher latitudes.

As opposed to sunspots, a **solar flare** generally disrupts radio communications because of increased ultraviolet and x-ray radiation. It takes about 8 minutes for this increased radiation to reach the Earth. An even more serious phenomenon is a **coronal mass ejection**, which causes charged particles to hit the Earth's atmosphere. After a coronal mass ejection, HF communications are disturbed. It takes about 20-40 hours for these particles to reach Earth and affect radio propagation.

High geomagnetic activity can cause **Aurora**, which can reflect VHF signals.

For one question on the test, you need to know that in the summer, there can be a lot of atmospheric static on the lower HF bands.

What is the significance of the sunspot number with regard to HF propagation?

A. Higher sunspot numbers generally indicate a greater probability of good propagation at higher frequencies

B. Lower sunspot numbers generally indicate greater probability of sporadic E propagation

C. A zero sunspot number indicate radio propagation is not possible on any band

D. All of these choices are correct.

What effect does a Sudden Ionospheric Disturbance have on the daytime ionospheric propagation of HF radio waves?

A. It enhances propagation on all HF frequencies

B. It disrupts signals on lower frequencies more than those on higher frequencies

C. It disrupts communications via satellite more than direct communications

D. None, because only areas on the night side of the Earth are affected

Approximately how long does it take the increased ultraviolet and X-ray radiation from solar flares to affect radio-wave propagation on the Earth?

A. 28 days

B. 1 to 2 hours

C. 8 minutes

D. 20 to 40 hours

Which of the following are least reliable for long distance communications during periods of low solar activity?

A. 80 meters and 160 meters

B. 60 meters and 40 meters

C. 30 meters and 20 meters

D. 15 meters, 12 meters and 10 meters

What is the solar flux index?

A. A measure of the highest frequency that is useful for ionospheric propagation between two points on the Earth

B. A count of sunspots which is adjusted for solar emissions

C. Another name for the American sunspot number

D. A measure of solar radiation at 10.7 centimeters wavelength

What is a geomagnetic storm?

A. A sudden drop in the solar-flux index

B. A thunderstorm which affects radio propagation

C. Ripples in the ionosphere

D. A temporary disturbance in the Earth's magnetosphere

At what point in the solar cycle does the 20 meter band usually support worldwide propagation during daylight hours?

A. At the summer solstice

B. Only at the maximum point of the solar cycle

C. Only at the minimum point of the solar cycle

D. At any point in the solar cycle

Which of the following effects can a geomagnetic storm have on radio propagation?

A. Improved high-latitude HF propagation

B. Degraded high-latitude HF propagation

C. Improved ground wave propagation

D. Degraded ground wave propagation

Which of the following is typical of the lower HF frequencies during the summer?

A. Poor propagation at any time of day

B. World-wide propagation during the daylight hours

C. Heavy distortion on signals due to photon absorption

D. High levels of atmospheric noise or "static"

What causes HF propagation conditions to vary periodically in a 28-day cycle?

A. Long term oscillations in the upper atmosphere

B. Cyclic variation in the Earth's radiation belts

C. The Sun's rotation on its axis

D. The position of the Moon in its orbit

What does the K-index indicate?

A. The relative position of sunspots on the surface of the Sun

B. The short term stability of the Earth's magnetic field

C. The stability of the Sun's magnetic field

D. The solar radio flux at Boulder, Colorado

Plain-English Study Guide for the FCC Amateur Radio General Class License

What does the A-index indicate?

A. The relative position of sunspots on the surface of the Sun

B. The amount of polarization of the Sun's electric field

C. The long term stability of the Earth's geomagnetic field

D. The solar radio flux at Boulder, Colorado

How are radio communications usually affected by the charged particles that reach the Earth from solar coronal holes?

A. HF communications are improved

B. HF communications are disturbed

C. VHF/UHF ducting is improved

D. VHF/UHF ducting is disturbed

How long does it take charged particles from coronal mass ejections to affect radio-wave propagation on the Earth?

A. 28 days

B. 14 days

C. 4 to 8 minutes

D. 20 to 40 hours

Richard P. Clem

What benefit can high geomagnetic activity have on radio communications?

A. Auroras that can reflect VHF signals

B. Higher signal strength for HF signals passing through the polar regions

C. Improved HF long path propagation

D. Reduced long delayed echoes

As you probably know, HF radio signals are reflected (or more correctly, refracted) by the **ionosphere**, and can travel a long distance around the Earth. Sometimes, the signal might travel two different paths. One is the **long path**, and the other is the **short path**. These signals arrive at a slightly different time, and there would be a definite echo that can be heard.

One interesting phenomenon that will indicate that a higher band might be "open" is the presence of short-skip signals on a lower band. For example, if you start hearing stations fairly close to you on 10 meters, this might indicate that 6 meters is open. Similarly, if you hear short skip on 6 meters, this might indicate that 2 meters is open.

Two important frequencies to know are the **MUF (Maximum Usable Frequency)** and **LUF (lowest usable frequency)**. These will vary based upon solar activity. As the name implies, the MUF is the highest frequency that can be used to support communication via the ionosphere. To put it another way, signals between these two frequencies are bent back by the ionosphere and can be heard over large distances. One interesting phenomenon is that frequencies very close to the MUF have less attenuation. So, for example, if the MUF is 29 MHz, then 28 MHz would have very low attenuation and very good operating conditions.

Plain-English Study Guide for the FCC Amateur Radio General Class License

MUF is affected by all of the following factors: path location and distance, time of day, season, solar radiation, and ionospheric disturbances. If the LUF ever gets higher than the MUF, then there is no HF frequency that can be used over the path.

Frequencies below the LUF are completely absorbed by the ionosphere and do not get bent back toward Earth.

HF signals are generally reflected by the **F2 layer** of the ionosphere, and one "hop" back to Earth generally covers about 2500 miles. On the higher HF frequencies and on VHF, most of this skip takes place from the **E region** of the ionosphere. One "hop" from the E layer is about 1200 miles.

These layers are named by letter, starting at the Earth's surface. Therefore, the layer with the lowest letter is closest to the Earth. The **D layer** is the first one that has an affect on radio communication. The D layer makes long-distance communications during the day on the lower frequencies (40-160 meters) difficult because it absorbs signals on those frequencies. The F2 layer is responsible for the longest distance propagation because it is the highest.

Radio propagation depends on the time of day. This is because the ionospheric layers reach their maximum height where the sun is overhead.

A good way to determine the MUF is to listen to beacon stations. If you can hear them, then the frequency of that station is below the MUF. International beacon stations operate on the following frequencies, and you should avoid these frequencies so that you do not interfere: 14.100, 18.110, 21.150, 24. 930 and 28.200 MHz

You need to know the term **critical angle** for the test. This is the highest takeoff angle that will return a radio wave to the Earth under specific ionospheric conditions.

Scatter signals on HF will have a fluttering sound because energy is scattered into the skip zone through several different radio wave paths. Scatter signals in the skip zone are usually weak because only a small part of the signal energy is scattered there. But scatter is often the only way to communicate in the **skip zone**--the area beyond the ground wave, but too close to receive sky wave signals. If signals are being heard above the MUF, this might be scatter propagation.

There is one question about the best antenna for skip communication on 40 meters during the day. The correct answer is horizontal dipoles between 1/8 and 1/4 wavelength above the ground. (The correct answer to another question is "between 1/10 and ¼ wavelength." Since these figures are approximate, either one is correct.) This is because such an antenna is good for **NVIS--Near Vertical Incidence Sky-wave** propagation. This is short distance HF propagation using high elevation angles. An NVIS antenna will have high vertical angle radiation. This is another way of saying that much of the signal will go straight up.

How might a sky-wave signal sound if it arrives at your receiver by both short path and long path propagation?

A. Periodic fading approximately every 10 seconds

B. Signal strength increased by 3 dB

C. The signal might be cancelled causing severe attenuation

D. A well-defined echo might be heard

Which of the following applies when selecting a frequency for lowest attenuation when transmitting on HF?

A. Select a frequency just below the MUF

B. Select a frequency just above the LUF

C. Select a frequency just below the critical frequency

D. Select a frequency just above the critical frequency

What is a reliable way to determine if the MUF is high enough to support skip propagation between your station and a distant location on frequencies between 14 and 30 MHz?

A. Listen for signals from an international beacon in the frequency range you plan to use

B. Send a series of dots on the band and listen for echoes from your signal

C. Check the strength of TV signals from Western Europe

D. Check the strength of signals in the MF AM broadcast band

Why should an amateur operator normally avoid transmitting on 14.100, 18.110, 21.150, 24.930 and 28.200 MHz?

A. A system of propagation beacon stations operates on those frequencies

B. A system of automatic digital stations operates on those frequencies

C. These frequencies are set aside for emergency operations

D. These frequencies are set aside for bulletins from the FCC

What usually happens to radio waves with frequencies below the Maximum Usable Frequency (MUF) and above the Lowest Usable Frequency (LUF) when they are sent into the ionosphere?

A. They are bent back to the Earth

B. They pass through the ionosphere

C. They are amplified by interaction with the ionosphere

D. They are bent and trapped in the ionosphere to circle the Earth

What usually happens to radio waves with frequencies below the Lowest Usable Frequency (LUF)?

A. They are bent back to the Earth

B. They pass through the ionosphere

C. They are completely absorbed by the ionosphere

D. They are bent and trapped in the ionosphere to circle the Earth

What does LUF stand for?

A. The Lowest Usable Frequency for communications between two points

B. The Longest Universal Function for communications between two points

C. The Lowest Usable Frequency during a 24 hour period

D. The Longest Universal Function during a 24 hour period

What does MUF stand for?

A. The Minimum Usable Frequency for communications between two points

B. The Maximum Usable Frequency for communications between two points

C. The Minimum Usable Frequency during a 24 hour period

D. The Maximum Usable Frequency during a 24 hour period

What is the approximate maximum distance along the Earth's surface that is normally covered in one hop using the F2 region?

A. 180 miles

B. 1,200 miles

C. 2,500 miles

D. 12,000 miles

What is the approximate maximum distance along the Earth's surface that is normally covered in one hop using the E region?

A. 180 miles

B. 1,200 miles

C. 2,500 miles

D. 12,000 miles

What happens to HF propagation when the Lowest Usable Frequency (LUF) exceeds the Maximum Usable Frequency (MUF)?

A. No HF radio frequency will support ordinary skywave communications over the path

B. HF communications over the path are enhanced

C. Double hop propagation along the path is more common

D. Propagation over the path on all HF frequencies is enhanced

What factors affect the MUF?

A. Path distance and location

B. Time of day and season

C. Solar radiation and ionospheric disturbances

D. All of these choices are correct

Which of the following ionospheric layers is closest to the surface of the Earth?

A. The D layer

B. The E layer

C. The F1 layer

D. The F2 layer

Where on the Earth do ionospheric layers reach their maximum height?

Plain-English Study Guide for the FCC Amateur Radio General Class License

A. Where the Sun is overhead

B. Where the Sun is on the opposite side of the Earth

C. Where the Sun is rising

D. Where the Sun has just set

Why is the F2 region mainly responsible for the longest distance radio wave propagation?

A. Because it is the densest ionospheric layer

B. Because of the Doppler effect

C. Because it is the highest ionospheric region

D. Because of meteor trails at that level

What does the term "critical angle" mean as used in radio wave propagation?

A. The long path azimuth of a distant station

B. The short path azimuth of a distant station

C. The lowest takeoff angle that will return a radio wave to the Earth under specific ionospheric conditions

D. The highest takeoff angle that will return a radio wave to the Earth under specific ionospheric conditions

Why is long distance communication on the 40-meter, 60-meter, 80-meter and 160-meter bands more difficult during the day?

A. The F layer absorbs signals at these frequencies during daylight hours

B. The F layer is unstable during daylight hours

C. The D layer absorbs signals at these frequencies during daylight hours

D. The E layer is unstable during daylight hours

What is a characteristic of HF scatter?

A. Phone signals have high intelligibility

B. Signals have a fluttering sound

C. There are very large, sudden swings in signal strength

D. Scatter propagation occurs only at night

What makes HF scatter signals often sound distorted?

A. The ionospheric layer involved is unstable

B. Ground waves are absorbing much of the signal

C. The E-region is not present

D. Energy is scattered into the skip zone through several different radio wave paths

Plain-English Study Guide for the FCC Amateur Radio General Class License

Why are HF scatter signals in the skip zone usually weak?

A. Only a small part of the signal energy is scattered into the skip zone

B. Signals are scattered from the magnetosphere which is not a good reflector

C. Propagation is through ground waves which absorb most of the signal energy

D. Propagations is through ducts in F region which absorb most of the energy

What type of propagation allows signals to be heard in the transmitting station's skip zone?

A. Faraday rotation

B. Scatter

C. Chordal hop

D. Short-path

Which ionospheric layer is the most absorbent of long skip signals during daylight hours on frequencies below 10 MHz?

A. The F2 layer

B. The F1 layer

C. The E layer

D. The D layer

63

What is Near Vertical Incidence Sky-wave (NVIS) propagation?

A. Propagation near the MUF

B. Short distance HF propagation using high elevation angles

C. Long path HF propagation at sunrise and sunset

D. Double hop propagation near the LUF

Which of the following antenna types will be most effective as a Near Vertical Incidence Skywave (NVIS) antenna for short-skip communications on 40 meters during the day?

A. A horizontal dipole placed between 1/10 and 1/4 wavelength above the ground

B. A vertical antenna placed between 1/4 and 1/2 wavelength above the ground

C. A left-hand circularly polarized antenna

D. A right-hand circularly polarized antenna

Plain-English Study Guide for the FCC Amateur Radio General Class License

8 FCC RULES

You should become familiar with the frequency limits on which you are allowed to operate. At the very least, you should have a copy near your operating position, even if you don't memorize the entire chart. On the exam, only the HF frequencies are covered. You do not need to learn the band limits for the VHF and UHF bands (since they were covered on the Technician exam). Here are the general class frequency privileges that you should know:

160 Meters:

1.800-2.000 MHz: CW, Phone, Image, RTTY/Data

80/75 Meters:

3.525-3.600 MHz: CW, RTTY/Data*

3.800-4.000 MHz: CW, Phone, Image

*Note: While data is allowed in this entire segment, the test asks where data is normally used, and the answer is 3570 – 3600 kHz.

Richard P. Clem

60 Meters:

Five Specific Channels between 5.332 MHz and 5.405 MHz.

Note: On 60 meters, if you are using any kind of antenna other than a dipole, then you need to keep a record of the antenna's gain.

40 Meters*:

7.025-7.125 MHz : CW, RTTY/Data

7.175-7.300 MHz:: CW, Phone, Image

* Note: These 40-meter frequencies apply to stations in ITU Region 2, which is basically the Western Hemisphere (North and South America). If you are outside of ITU Region 2 (for example, in Guam), then different frequency ranges apply. On any band, if you are outside of Region 2, it's always a good idea to check, since the frequency bands might be different.

30 Meters:

10.100-10.150 MHz: CW, RTTY/Data

20 Meters:

14.025 -14.150 MHz CW, RTTY/Data*

14.225 -14.350 MHz: CW, Phone, Image

*Note: While data is allowed in this entire segment, the test asks where data is normally used, and the answer is 14.070 - 14.100 MHz. Most PSK-31 operation takes place at the bottom of this segment, near 14.070 MHz.

Plain-English Study Guide for the FCC Amateur Radio General Class License

17 Meters:

18.068-18.110 MHz: CW, RTTY/Data

18.110-18.168 MHz: CW, Phone, Image

15 Meters:

21.025-21.200 MHz: CW, RTTY/Data

21.275-21.450 MHz: CW, Phone, Image

12 Meters:

24.890-24.930 MHz: CW, RTTY/Data

24.930-24.990 MHz: CW, Phone, Image

10 Meters:

28.000-28.300 MHz: CW, RTTY/Data

28.300-29.700 MHz: CW, Phone, Image

The top portion of the 10 meter band (29.5 – 29.7 MHz) is available for repeater use.

Even if you don't memorize the entire chart, the following facts will allow you to answer most of the questions correctly: First of all, the bands are usually identified by their wavelength on the exam, and some of the answers are clearly wrong. If the question asks you to identify a frequency on the 10 meter band, then 28.600 MHz might be a correct answer. But if there is an answer of 14.320, that answer is clearly wrong. Remember, frequency (in MHz) times wavelength (in meters) will equal approximately 300. This is often enough information to answer all of the frequency questions.

$f = \dfrac{300}{\lambda} \Longleftarrow 300 = f(\lambda) \Longrightarrow \lambda = \dfrac{300}{f}$

Also, keep in mind that the CW/Digital portion of the band is usually at the bottom of the band, and the phone portion of the band is usually at the top. And on those bands where General class licensees are not permitted to use phone on the entire phone band, the General portion of the band is usually at the top of the band.

Also, keep the following facts in mind: Phone or image transmissions are not allowed on 30 meters. That band can be used for CW and data only.

On 160, 60, 30, 17, 12, and 10 meters, General class licensees can use the entire band. On the other HF bands (80/75, 40, 20, and 15), there are segments of the bands reserved only for higher class licensees.

60 meters is the only band where Amateurs are allowed only on specific channels. On other bands, Amateurs may transmit anywhere within the band limits. The maximum permitted bandwidth on 60 meters is 2.8 kHz.

In some cases (for example, 60 or 30 meters), the Amateur Service is a **secondary user** of a band. That means that there are other non-Amateur stations that are allowed to use the band, and that we share it with them. If we are secondary, then we may use the band only if we do not cause interference to the primary users. If a primary user interferes with you on these bands, then you need to move to a clear frequency.

In other cases, the Amateur Service shares a band with another user, but we are primary, meaning that we have priority. For example, the 2.4 GHz band is shared with unlicensed Wi-Fi devices. However, an amateur station is **not** allowed to communicated with these unlicensed devices.

No Amateur band is shared with the Citizens Radio Service. This is a trick question.

Plain-English Study Guide for the FCC Amateur Radio General Class License

In addition to the FCC rules covering frequencies, there are voluntary band plans that amateurs use. When selecting a frequency, it is good amateur practice to follow the band plan for the mode you are using.

On some bands, the band plan includes a "DX window." This is the part of the band that should not be used for contacts within the 48 contiguous United States. On 6 meters, the DX window is 50.1 to 50.125, and stations in the 48 states should use it only for contacts outside the 48 contiguous states.

In general, amateurs are not allowed to make one-way transmissions. There are a few exceptions, and one of them is on the test. Amateurs are allowed to make one-way transmissions to help people learn Morse code. For the test, you also need to know that it is OK to occasionally retransmit weather and propagation information from U.S. Government statons.

On which HF/MF bands is a General class license holder granted all amateur frequency privileges?

A. 60 meters, 20 meters, 17 meters, and 12 meters

B. 160 meters, 80 meters, 40 meters, and 10 meters

C. 160 meters, 60 meters, 30 meters, 17 meters, 12 meters, and 10 meters

D. 160 meters, 30 meters, 17 meters, 15 meters, 12 meters, and 10 meters

On which of the following bands is phone operation prohibited?

A. 160 meters

B. 30 meters

C. 17 meters

D. 12 meters

On which of the following bands is image transmission prohibited?

A. 160 meters

B. 30 meters

C. 20 meters

D. 12 meters

Which of the following amateur bands is restricted to communication on only specific channels, rather than frequency ranges?

A. 11 meters

B. 12 meters

C. 30 meters

D. 60 meters

Which of the following frequencies is in the General class portion of the 40-meter band in ITU Region 2?

A. 7.250 MHz

B. 7.500 MHz

C. 40.200 MHz

D. 40.500 MHz

Plain-English Study Guide for the FCC Amateur Radio General Class License

Which of the following frequencies is within the General Class portion of the 75 meter phone band?

A. 1875 kHz

B. 3750 kHz

C. 3900 kHz

D. 4005 kHz

Which of the following frequencies is within the General Class portion of the 20 meter phone band?

A. 14005 kHz

B. 14105 kHz

C. 14305 kHz

D. 14405 kHz

Which of the following frequencies is within the General Class portion of the 80 meter band?

A. 1855 kHz

B. 2560 kHz

C. 3560 kHz

D. 3650 kHz

Which of the following frequencies is within the General Class portion of the 15 meter band?

A. 14250 kHz

B. 18155 kHz

C. 21300 kHz

D. 24900 kHz

Which of the following frequencies is available to a control operator holding a General Class license?

A. 28.020 MHz

B. 28.350 MHz

C. 28.550 MHz

D. All of these choices are correct

When General Class licensees are not permitted to use the entire voice portion of a particular band, which portion of the voice segment is generally available to them?

A. The lower frequency end

B. The upper frequency end

C. The lower frequency end on frequencies below 7.3 MHz and the upper end on frequencies above 14.150 MHz

D. The upper frequency end on frequencies below 7.3 MHz and the lower end on frequencies above 14.150 MHz

Plain-English Study Guide for the FCC Amateur Radio General Class License

Which of the following applies when the FCC rules designate the Amateur Service as a secondary user on a band?

A. Amateur stations must record the call sign of the primary service station before operating on a frequency assigned to that station

B. Amateur stations are allowed to use the band only during emergencies

C. Amateur stations are allowed to use the band only if they do not cause harmful interference to primary users

D. Amateur stations may only operate during specific hours of the day, while primary users are permitted 24 hour use of the band

What is the appropriate action if, when operating on either the 30-meter or 60-meter bands, a station in the primary service interferes with your contact?

A. Notify the FCCs regional Engineer in Charge of the interference

B. Increase your transmitter's power to overcome the interference

C. Attempt to contact the station and request that it stop the interference

D. Move to a clear frequency or stop transmitting

Which of the following may apply in areas under FCC jurisdiction outside of ITU Region 2?

A. Station identification may have to be in a language other than English

B. Morse code may not be permitted

C. Digital transmission may not be permitted

D. Frequency allocations may differ

What portion of the 10-meter band is available for repeater use?

A. The entire band

B. The portion between 28.1 MHz and 28.2 MHz

C. The portion between 28.3 MHz and 28.5 MHz

D. The portion above 29.5 MHz

What is the maximum bandwidth permitted by FCC rules for Amateur Radio stations when transmitting on USB frequencies in the 60 meter band?

A. 2.8 kHz

B. 5.6 kHz

C. 1.8 kHz

D. 3 kHz

The frequency allocations of which ITU region apply to radio amateurs operating in North and South America?

A. Region 4

B. Region 3

C. Region 2

D. Region 1

What segment of the 20-meter band is most often used for digital transmissions (avoiding the DX propagation beacons)?

A. 14.000 - 14.050 MHz

B. 14.070 - 14.112 MHz

C. 14.150 - 14.225 MHz

D. 14.275 - 14.350 MHz

What segment of the 80-meter band is most commonly used for digital transmissions?

A. 3570 – 3600 kHz

B. 3500 – 3525 kHz

C. 3700 – 3750 kHz

D. 3775 – 3825 kHz

In what segment of the 20 meter band are most PSK31 operations commonly found?

A. At the bottom of the slow-scan TV segment, near 14.230 MHz

B. At the top of the SSB phone segment near 14.325 MHz

C. In the middle of the CW segment, near 14.100 MHz

D. Below the RTTY segment, near 14.070 MHz

Which of the following complies with good amateur practice when choosing a frequency on which to initiate a call?

A. Check to see if the channel is assigned to another station

B. Identify your station by transmitting your call sign at least 3 times

C. Follow the voluntary band plan for the operating mode you intend to use

D. All of these choices are correct

What is the voluntary band plan restriction for U.S. stations transmitting within the 48 contiguous states in the 50.1 to 50.125 MHz band segment?

A. Only contacts with stations not within the 48 contiguous states

B. Only contacts with other stations within the 48 contiguous states

C. Only digital contacts

D. Only SSTV contacts

Which of the following is required by the FCC rules when operating in the 60-meter band?

A. If you are using other than a dipole antenna, you must keep a record of the gain of your antenna

B. You must keep a record of the date, time, frequency, power level and stations worked

C. You must keep a record of all third party traffic

D. You must keep a record of the manufacturer of your equipment and the antenna used

Which of the following one-way transmissions are permitted?

A. Unidentified test transmissions of less than one minute in duration

B. Transmissions necessary to assist learning the International Morse code

C. Regular transmissions offering equipment for sale, if intended for Amateur Radio use

D. All these choices are correct

In what part of the 13-centimeter band may an amateur station communicate with non-licensed Wi-Fi stations?

A. Anywhere in the band

B. Channels 1 through 4

C. Channels 42 through 45

D. No part

On what band do amateurs share channels with the unlicensed Wi-Fi service?

A. 432 MHz

B. 902 MHz

C. 2.4 GHz

D. 10.7 GHz

Richard P. Clem

Which of the following transmissions is permitted?

A. Unidentified transmissions for test purposes only

B. Retransmission of other amateur station signals by any amateur station

C. Occasional retransmission of weather and propagation forecast information from U.S. government stations

D. Coded messages of any kind, if not intended to facilitate a criminal act

There are some other rules that are covered on the test. The following is not an exhaustive listing of all FCC rules. This is just a summary of the rules that are on the test. In addition to reading this study guide, you should occasionally sit down and read the rules. And even if the rules do not have a specific provision about something, you are required to use good engineering and good amateur practice. "Good amateur practice" is determined by the FCC.

Unless your antenna is near an airport, you do not need to notify the FAA of your antenna, unless it is over 200 feet above ground level.

An amateur may operate a beacon station in order to observe propagation and reception. Automatic beacons are only allowed on certain frequencies. For the test, you need to know that they are allowed between 28.20 and 28.30 MHz. You are not allowed to have more than one beacon signal on the same band from a single location. The power limit for beacons is 100 watts.

In general, secret codes may not be used on the air. However, abbreviations and procedural signals may be used on the air, as long as they do not obscure the meaning of the message. Before using a new digital protocol on the air, it is necessary to publicly document its technical characteristics.

When choosing your transmitting frequency, you should do all of the following:

1. Review FCC Part 97 Rules regarding permitted frequencies and emissions?

2. Follow generally accepted band plans agreed to by the Amateur Radio community.

3. Before transmitting, listen to avoid interfering with ongoing communication

In general, the transmitting power limit is 1500 watts PEP output. (The FCC always uses PEP to specify power levels. However, there are some exceptions. For example, on the 30 meter band (10.1 - 10.15 MHz), the limit is 200 watts. The limit on the 60 meter band is 100 watts, but has some special wording. You need to remember that it is 100 watts ERP "with respect to a dipole." The limit for spread spectrum transmissions is 10 watts. Also, you should always use the minimum amount of power necessary to carry out the desired communications.

The maximum symbol rate permitted for RTTY or data emission below 28 MHz is 300 baud. On the 10 meter band, it is 1200 baud. On 2 meters, it is 19.6 kilobaud. On 222 MHz, it is 56 kilobaud.

One question that is on the test is whether it is legal for a repeater to retransmit a Technician's 2-meter signal on 10-meters. This is a question because the Technician is not allowed to transmit on the repeater portion of the 10 meter band. The answer to this question is that it is legal to do so (provided, of course, that the control operator of the repeater has at least a General license).

There are a few questions on the test about interference, although they do not go into as much detail as the Technician exam.

To avoid interference, common sense is usually the best guide. No one has priority access to frequencies. During a contact, propagation conditions sometimes change, and you might notice increasing interference from other stations you didn't hear before. If this happens, it is best to use common courtesy and attempt to resolve the problem in a mutually acceptable manner. This might include your moving to another frequency.

If you are in communication with someone and another station in distress breaks in, then you should obviously acknowledge the station and determine what assistance might be needed. During an actual emergency, you can use any means at your disposal to assist another station in distress. And obviously, if you need to send a distress call, then you should use whatever frequency has the best chance of communicating the distress message.

If there is interference between a coordinated repeater and an uncoordinated repeater, then the licensee of the non-coordinated repeater has the primary responsibility to resolve the interference.

There are special interference rules that apply in the following three situations: 1. If you are operating within one mile of an FCC Monitoring Station; 2. If you are using a band where the Amateur Service is secondary; 3. If you are transmitting spread spectrum emissions. You don't need to know those special rules for the test, but you do need to be aware that there are rules for those three situations.

The rules covering identifying your station were covered on the Technician exam. For the General exam, you also need to know that unidentified transmissions are not permitted, even for tests, and that you must identify in English.

The **Amateur Auxiliary** to the FCC is a group of amateur volunteers who are formally enlisted to monitor the airwaves for rules violations. Their objective is to encourage amateur self regulation and compliance with the rules. Hidden transmitter hunts are helpful to the Amateur Auxiliary because they provide practice in direction finding used to locate stations violating the rules.

Plain-English Study Guide for the FCC Amateur Radio General Class License

The test asks you whether it is permissible to communicate with amateur stations in other countries (in other words, "outside areas administered by the FCC.") You may, unless that country has notified the ITU that they object to such communications. (As of the time of writing, no country has made this objection.) Note: Don't confuse this rule with the rules for third-party traffic with other countries. Third-party traffic is permitted **only** if the other country has an agreement with the U.S. to permit it. But a normal contact is always permitted, **unless** the other country objects.

There are two questions on the test about the rules governing **RACES (Radio Amateur Civil Emergency Service)**. For the test, you need to know that the control operator of a RACES station must have an FCC amateur radio license. You also need to know that the FCC may restrict normal frequency operations of stations participating in RACES when the President's War Emergency Powers have been invoked.

Automatic control of a station is allowed only on certain frequencies. You don't need to know the exact frequencies for the test, but you need to know that they are certain frequencies on 80-2 meters, and all frequencies in the 6 meter band and above). For example, an unattended digital station that transfers messages to and from the Internet would be considered an automatically controlled station. If you are outside the band segments set aside for automatic control, then the station initiating the contact must be under local or remote control.

There is one question on the test about state and local antenna rules. You need to know that these laws must reasonably accommodate amateur radio and must be the minimum restriction that is practical.

What is the maximum height above ground to which an antenna structure may be erected without requiring notification to the FAA and registration with the FCC, provided it is not at or near a public use airport?

A. 50 feet

B. 100 feet

C. 200 feet

D. 300 feet

With which of the following conditions must beacon stations comply?

A. A beacon station may not use automatic control

B. The frequency must be coordinated with the National Beacon Organization

C. The frequency must be posted on the internet or published in a national periodical

D. There must be no more than one beacon signal transmitting in the same band from the same station location

On what HF frequencies are automatically controlled beacons permitted?

A. On any frequency if power is less than 1 watt

B. On any frequency if transmissions are in Morse code

C. 21.08 MHz to 21.09 MHz

D. 28.20 MHz to 28.30 MHz

Which of the following is a purpose of a beacon station as identified in the FCC Rules?

A. Observation of propagation and reception

Plain-English Study Guide for the FCC Amateur Radio General Class License

B. Automatic identification of repeaters

C. Transmission of bulletins of general interest to Amateur Radio licensees

D. Identifying net frequencies

What must be done before using a new digital protocol on the air?

A. Type-certify equipment to FCC standards

B. Obtain an experimental license from the FCC

C. Publicly document the technical characteristics of the protocol

D. Submit a rule-making proposal to the FCC describing the codes and methods of the technique

What are the restrictions on the use of abbreviations or procedural signals in the Amateur Service?

A. Only "Q" codes are permitted

B. They may be used if they do not obscure the meaning of a message

C. They are not permitted

D. Only "10 codes" are permitted

When choosing a transmitting frequency, what should you do to comply with good amateur practice?

A. Insure that the frequency and mode selected are within your license class privileges

B. Follow generally accepted band plans agreed to by the Amateur Radio community

C. Monitor the frequency before transmitting

D. All of these choices are correct

What measurement is specified by FCC rules that regulate maximum power output?

A. RMS

B. Average

C. Forward

D. PEP

What is the power limit for beacon stations?

A. 10 watts PEP output

B. 20 watts PEP output

C. 100 watts PEP output

D. 200 watts PEP output

What is the maximum power limit on the 60-meter band?

A. 1500 watts PEP

B. 10 watts RMS

Plain-English Study Guide for the FCC Amateur Radio General Class License

C. ERP of 100 watts PEP with respect to a dipole

D. ERP of 100 watts PEP with respect to an isotropic antenna

What is the maximum PEP output allowed for spread spectrum transmissions?

A. 100 milliwatts

B. 10 watts

C. 100 watts

D. 1500 watts

Who or what determines "good engineering and good amateur practice" as applied to the operation of an amateur station in all respects not covered by the Part 97 rules?

A. The FCC

B. The Control Operator

C. The IEEE

D. The ITU

What is the maximum transmitting power an amateur station may use on 10.140 MHz?

A. 200 watts PEP output

B. 1000 watts PEP output

C. 1500 watts PEP output

D. 2000 watts PEP output

G1C02 (C) [97.313]

What is the maximum transmitting power an amateur station may use on the 12-meter band?

A. 50 watts PEP output

B. 200 watts PEP output

C. 1500 watts PEP output

D. An effective radiated power equivalent to 100 watts from a half-wave dipole

Which of the following limitations apply to transmitter power on every amateur band?

A. Only the minimum power necessary to carry out the desired communications should be used

B. Power must be limited to 200 watts when using data transmissions

C. Power should be limited as necessary to avoid interference to another radio service on the frequency

D. Effective radiated power cannot exceed 1500 watts

Which of the following is a limitation on transmitter power on the 28 MHz band for a General Class control operator?

A. 100 watts PEP output

B. 1000 watts PEP output

C. 1500 watts PEP output

D. 2000 watts PEP output

Which of the following is a limitation on transmitter power on 1.8 MHz band?

A. 200 watts PEP output

B. 1000 watts PEP output

C. 1200 watts PEP output

D. 1500 watts PEP output

What is the maximum symbol rate permitted for RTTY or data emission transmission on the 20 meter band?

A. 56 kilobaud

B. 19.6 kilobaud

C. 1200 baud

D. 300 baud

What is the maximum symbol rate permitted for RTTY or data emission transmitted at frequencies below 28 MHz?

A. 56 kilobaud

B. 19.6 kilobaud

C. 1200 baud

D. 300 baud

What is the maximum symbol rate permitted for RTTY or data emission transmitted on the 1.25 meter and 70 centimeter bands

A. 56 kilobaud

B. 19.6 kilobaud

C. 1200 baud

D. 300 baud

What is the maximum symbol rate permitted for RTTY or data emission transmissions on the 10 meter band?

A. 56 kilobaud

B. 19.6 kilobaud

C. 1200 baud

D. 300 baud

What is the maximum symbol rate permitted for RTTY or data emission transmissions on the 2 meter band?

A. 56 kilobaud

B. 19.6 kilobaud

C. 1200 baud

D. 300 baud

When may a 10-meter repeater retransmit the 2-meter signal from a station having a Technician Class control operator?

A. Under no circumstances

B. Only if the station on 10-meters is operating under a Special Temporary Authorization allowing such retransmission

C. Only during an FCC declared general state of communications emergency

D. Only if the 10-meter repeater control operator holds at least a General Class license

Which of the following conditions require an Amateur Radio station licensee to take specific steps to avoid harmful interference to other users or facilities?

A. When operating within one mile of an FCC Monitoring Station

B. When using a band where the Amateur Service is secondary

C. When a station is transmitting spread spectrum emissions

D. All of these choices are correct

Which of the following is true concerning access to frequencies?

A. Nets always have priority

B. QSOs in progress always have priority

C. Except during emergencies, no amateur station has priority access to any frequency

D. Contest operations must always yield to non-contest use of frequencies

What is the first thing you should do if you are communicating with another amateur station and hear a station in distress break in?

A. Continue your communication because you were on frequency first

B. Acknowledge the station in distress and determine what assistance may be needed

C. Change to a different frequency

D. Immediately cease all transmissions

What is good amateur practice if propagation changes during a contact and you notice interference from other stations on the frequency?

A. Tell the interfering stations to change frequency

B. Report the interference to your local Amateur Auxiliary Coordinator

C. Attempt to resolve the interference problem with the other stations in a mutually acceptable manner

D. Increase power to overcome interference

What frequency should be used to send a distress call?

A. Whatever frequency has the best chance of communicating the distress message

B. Only frequencies authorized for RACES or ARES stations

C. Only frequencies that are within your operating privileges

D. Only frequencies used by police, fire or emergency medical services

When is an amateur station allowed to use any means at its disposal to assist another station in distress?

A. Only when transmitting in RACES

B. At any time when transmitting in an organized net

C. At any time during an actual emergency

D. Only on authorized HF frequencies

When is it permissible to communicate with amateur stations in countries outside the areas administered by the Federal Communications Commission?

A. Only when the foreign country has a formal third party agreement filed with the FCC

B. When the contact is with amateurs in any country except those whose administrations have notified the ITU that they object to such communications

C. When the contact is with amateurs in any country as long as the communication is conducted in English

D. Only when the foreign country is a member of the International Amateur Radio Union

Who may be the control operator of an amateur station transmitting in RACES to assist relief operations during a disaster?

A. Only a person holding an FCC issued amateur operator license

B. Only a RACES net control operator

C. A person holding an FCC issued amateur operator license or an appropriate government official

D. Any control operator when normal communication systems are operational

What is required to conduct communications with a digital station operating under automatic control outside the automatic control band segments?

A. The station initiating the contact must be under local or remote control

B. The interrogating transmission must be made by another automatically controlled station

C. No third party traffic maybe be transmitted

D. The control operator of the interrogating station must hold an Extra Class license

On what bands may automatically controlled stations transmitting RTTY or data emissions communicate with other automatically controlled digital station?

Plain-English Study Guide for the FCC Amateur Radio General Class License

A. On any band segment where digital operation is permitted

B. Anywhere in the non-phone segments of the 10-meter or shorter wavelength bands

C. Only in the non-phone Extra Class segments of the bands

D. Anywhere in the 6-meter or shorter wavelength bands, and in limited segments of some of the HF bands

Under what conditions are state and local governments permitted to regulate Amateur Radio antenna structures?

A. Under no circumstances, FCC rules take priority

B. At any time and to any extent necessary to accomplish a legitimate purpose of the state or local entity, provided that proper filings are made with the FCC

C. Only when such structures exceed 50 feet in height and are clearly visible 1000 feet from the structure

D. Amateur Service communications must be reasonably accommodated, and regulations must constitute the minimum practical to accommodate a legitimate purpose of the state or local entity

The exam also covers some of the rules regarding volunteer examiners. When you take any amateur exam, it is given by a **volunteer examiner (VE)**. When you get your General class license, you will be eligible to become a VE. If you become a VE as a general class licensee, then you are eligible to administer only the Technician class exam.

After you pass the test, you can immediately begin operating on general class frequencies, as soon as you receive your CSCE (Certificate of Successful Completion of Element). On the air, when using general frequencies, you need to identify your call sign followed by "slant AG". A CSCE is valid for 365 days (one year).

Volunteer Examiners are accredited by an organization known as a **Volunteer Examiner Coordinator (VEC)**. You must be accredited by a VEC before you can give an exam. Three VE's are required in order to give any exam. In the case of a technician exam, the three VE's can be general class or higher.

A VE must be at least 18 years old. A VE is **not** required to be a U.S. Citizen.

There are two questions about getting re-licensed after an old license has expired. Anyone who previously held a General, Advanced, or Extra class license can get credit for those elements, but they must pass the current element 2 exam.

When must you add the special identifier "AG" after your call sign if you are a Technician Class licensee and have a CSCE for General Class operator privileges, but the FCC has not yet posted your upgrade on its website?

A. Whenever you operate using General Class frequency privileges

B. Whenever you operate on any amateur frequency

C. Whenever you operate using Technician frequency privileges

D. A special identifier is not required as long as your General Class license application has been filed with the FCC

Plain-English Study Guide for the FCC Amateur Radio General Class License

What license examinations may you administer when you are an accredited VE holding a General class operator license?

A. General and Technician

B. General only

C. Technician only

D. Amateur Extra, General, and Technician

On which of the following band segments may you operate if you are a Technician Class operator and have a CSCE for General Class privileges?

A. Only the Technician band segments until your upgrade is posted on the FCC database

B. Only on the Technician band segments until your license arrives in the mail

C. On any General or Technician Class band segment

D. On any General or Technician Class band segment except 30 and 60 meters

Which of the following is a requirement for administering a Technician Class license examination?

A. At least three General Class or higher VEs must observe the examination

B. At least two General Class or higher VEs must be present

C. At least two General Class or higher VEs must be present, but only one need be Extra Class

D. At least three VEs of Technician Class or higher must observe the examination

Which of the following must a person have before they can be an administering VE for a Technician Class license examination?

A. Notification to the FCC that you want to give an examination

B. Receipt of a CSCE for General Class

C. Possession of a properly obtained telegraphy license

D. An FCC General Class or higher license and VEC accreditation

Volunteer Examiners are accredited by what organization?

A. The Federal Communications Commission

B. The Universal Licensing System

C. A Volunteer Examiner Coordinator

D. The Wireless Telecommunications Bureau

Which of the following criteria must be met for a non-U.S. citizen to be an accredited Volunteer Examiner?

A. The person must be a resident of the U.S. for a minimum of 5 years

B. The person must hold an FCC granted Amateur Radio license of General Class or above

C. The person's home citizenship must be in the ITU 2 region

D. None of these choices is correct; non-U.S. citizens cannot be volunteer examiners

How long is a Certificate of Successful Completion of Examination (CSCE) valid for exam element credit?

A. 30 days

B. 180 days

C. 365 days

D. For as long as your current license is valid

What is the minimum age that one must be to qualify as an accredited Volunteer Examiner?

A. 12 years

B. 18 years

C. 21 years

D. There is no age limit

Who may receive credit for the elements represented by an expired amateur radio license?

Richard P. Clem

A. Any person who can demonstrate that they once held an FCC issued General, Advanced, or Amateur Extra class license that was not revoked by the FCC

B. Anyone who held an FCC issued amateur radio license that has been expired for not less than 5 years and not more than 15 years

C. Any person who previously held an amateur license issued by another country, but only if that country has a current reciprocal licensing agreement with the FCC

D. Only persons who once held an FCC issued Novice, Technician, or Technician Plus license

What is required to obtain a new General Class license after a previously-held license has expired and the two-year grace period has passed?

A. They must have a letter from the FCC showing they once held an amateur or commercial license

B. There are no requirements other than being able to show a copy of the expired license

C. The applicant must be able to produce a copy of a page from a call book published in the U.S. showing his or her name and address

D. The applicant must pass the current Element 2 exam

As an amateur, you are permitted to send messages on behalf of other people, and this includes allowing another person to talk into the microphone of your station. There are a number of restrictions, however; and some of these are covered on the test.

A person may not participate in stating a message over your station if that person once held an amateur license that was revoked. And you may not

Plain-English Study Guide for the FCC Amateur Radio General Class License

transmit messages to or from third parties if you are contacting a station in another country, unless that country has an agreement permitting such messages. This is called a **third party agreement**. (This would include allowing another person to speak into your microphone while making a "DX" contact: That is not permitted unless the United States has a third party agreement with the other country.) In addition, such messages to or from a foreign country must messages relating to Amateur Radio or be remarks of a personal character. Or, they can be messages relating to emergencies or disaster relief.

Which of the following would disqualify a third party from participating in stating a message over an amateur station?

A. The third party's amateur license has been revoked and not reinstated

B. The third party is not a U.S. citizen

C. The third party is a licensed amateur

D. The third party is speaking in a language other than English

What types of messages for a third party in another country may be transmitted by an amateur station?

A. Any message, as long as the amateur operator is not paid

B. Only messages for other licensed amateurs

C. Only messages relating to Amateur Radio or remarks of a personal character, or messages relating to emergencies or disaster relief

D. Any messages, as long as the text of the message is recorded in the station log

Richard P. Clem

9 OPERATING PROCEDURES

The test has a number of questions about operating practices and procedures, in addition to what is required by the rules.

Many amateurs keep a log book. The station log usually contains the date and time of contact, the band and/or frequency, the other station's call sign, and the signal report. Even though station logs are generally not required by the FCC, many amateurs still keep one. One good reason for doing so is in case the FCC requests information about your station.

When using voice, the FCC rules encourage the use of a standard phonetic alphabet, although the rules themselves don't require a certain standard. The most common, however, originally adopted by the International Civil Aviation Organization, is frequently called the NATO phonetic alphabet. For the test, you need to know that the first four letters are represented by Alpha, Bravo, Charlie, and Delta. (If you remember Alpha, Bravo, **or** Delta, then you'll get the question right by process of elimination.)

There's one trick question on the test about contests. There are no special requirements about contests. You just need to remember that the normal rules apply, including identifying as usual.

Plain-English Study Guide for the FCC Amateur Radio General Class License

There are a few questions on the test about **CW** (Morse Code) operating procedures.

When selecting a CW transmitting frequency, the minimum frequency separation to avoid interference with other stations is about 150 to 500 Hz.

When answering a CQ in Morse Code, it is best to use the same speed at which the CQ was sent.

In CW operation, "zero beat" means matching your transmit frequency exactly to the frequency of the received signal.

The **RST** system (Readability, Signal Strength, Tone) is used to give signal reports. When using CW, the letter "C" at the end of the report means that the other station has a chirpy or unstable signal. The best way to check the waveform of your transmitted signal (to make sure it's not chirping) is to use an oscilloscope.

When receiving CW, it is sometimes useful to use the opposite or "reverse" sideband. By doing this, it might be possible to reduce or eliminate interference from other stations.

An electronic keyer is a device for automatically generating strings of dots and dashes for CW operation.

What is a reason why many amateurs keep a station log?

A. The ITU requires a log of all international contacts

B. The ITU requires a log of all international third party traffic

C. The log provides evidence of operation needed to renew a license without retest

D. To help with a reply if the FCC requests information

Richard P. Clem

Which of the following are examples of the NATO Phonetic Alphabet?

A. Able, Baker, Charlie, Dog

B. Adam, Boy, Charles, David

C. America, Boston, Canada, Denmark

D. Alpha, Bravo, Charlie, Delta

Which of the following is required when participating in a contest on HF frequencies?

A. Submit a log to the contest sponsor

B. Send a QSL card to the stations worked, or QSL via Logbook of The World

C. Identify your station per normal FCC regulations

D. All these choices are correct

When selecting a CW transmitting frequency, what minimum frequency separation should you allow in order to minimize interference to stations on adjacent frequencies?

A. 5 to 50 Hz

B. 150 to 500 Hz

C. 1 to 3 kHz

D. 3 to 6 kHz

Plain-English Study Guide for the FCC Amateur Radio General Class License

What is the best speed to use when answering a CQ in Morse code?

A. The fastest speed at which you are comfortable copying, but no slower than the CQ

B. The fastest speed at which you are comfortable copying, but no faster than the CQ

C. At the standard calling speed of 10 wpm

D. At the standard calling speed of 5 wpm

What does the term "zero beat" mean in CW operation?

A. Matching the speed of the transmitting station

B. Operating split to avoid interference on frequency

C. Sending without error

D. Matching your transmit frequency to the frequency of a received signal.

When sending CW, what does a "C" mean when added to the RST report?

A. Chirpy or unstable signal

B. Report was read from S meter reading rather than estimated

C. 100 percent copy

D. Key clicks

What is one advantage of selecting the opposite or "reverse" sideband when receiving CW signals on a typical HF transceiver?

A. Interference from impulse noise will be eliminated

B. More stations can be accommodated within a given signal passband

C. It may be possible to reduce or eliminate interference from other signals

D. Accidental out of band operation can be prevented

What is the purpose of an electronic keyer?

A. Automatic transmit/receive switching

B. Automatic generation of strings of dots and dashes for CW operation

C. VOX operation

D. Computer interface for PSK and RTTY operation

Which of the following is the best instrument to use when checking the keying waveform of a CW transmitter?

A. An oscilloscope

B. A field-strength meter

C. A sidetone monitor

D. A wavemeter

Plain-English Study Guide for the FCC Amateur Radio General Class License

10 ABBREVIATIONS AND Q-SIGNALS

The following abbreviations and Q-signals are covered on the test:

AR: End of message

CL: At the end of a transmission, this means that you are closing your station.

CQ: A general call to any station. Using voice, you would repeat "CQ" as few times, followed by "this is," followed by your call sign a few times.

CQ DX: The caller is looking for a station outside of their own country. For stations in the lower 48, this means stations outside the lower 48.

KN: At the end of the transmission, this means that you are listening only for a specific station or stations

MMIC: Monolithic Microwave Integrated Circuit

QRL?: Is the frequency in use? (Using QRL, or asking whether the frequency is in use, is a way to avoid causing harmful interference.) (It can also mean "are you busy?"

Richard P. Clem

QRP: Low power transmit operation

QRN: Static

QRS: Send slower

QRV: I am ready to receive messages

QSK: Full break-in telegraphy. This means that transmitting stations are able to receive between code characters and elements.

QSL: I acknowledge receipt

ROM: Read Only Memory

RTTY: Radioteletype

What does the Q signal "QRL?" mean?

A. "Will you keep the frequency clear?"

B. "Are you operating full break-in" or "Can you operate full break-in?"

C. "Are you listening only for a specific station?"

D. "Are you busy?", or "Is this frequency in use?"

Which of the following describes full break-in telegraphy (QSK)?

A. Breaking stations send the Morse code prosign BK

B. Automatic keyers are used to send Morse code instead of hand keys

C. An operator must activate a manual send/receive switch before and after every transmission

D. Transmitting stations can receive between code characters and elements

What should you do if a CW station sends "QRS"?

A. Send slower

B. Change frequency

C. Increase your power

D. Repeat everything twice

Generally, who should respond to a station in the contiguous 48 states who calls "CQ DX"?

A. Any caller is welcome to respond

B. Only stations in Germany

C. Any stations outside the lower 48 states

D. Only contest stations

What is a practical way to avoid harmful interference when selecting a frequency to call CQ on CW or phone?

A. Send "QRL?" on CW, followed by your call sign; or, if using phone, ask if the frequency is in use, followed by your call sign

B. Listen for 2 minutes before calling CQ

C. Send the letter "V" in Morse code several times and listen for a response

D. Send "QSY" on CW or if using phone, announce "the frequency is in use", then send your call and listen for a response

Which of the following is a good way to indicate on a clear frequency in the HF phone bands that you are looking for a contact with any station?

A. Sign your call sign once, followed by the words "listening for a call" -- if no answer, change frequency and repeat

B. Say "QTC" followed by "this is" and your call sign -- if no answer, change frequency and repeat

C. Repeat "CQ" a few times, followed by "this is," then your call sign a few times, then pause to listen, repeat as necessary

D. Transmit an unmodulated carried for approximately 10 seconds, followed by "this is" and your call sign, and pause to listen -- repeat as necessary

What does it mean when a CW operator sends "KN" at the end of a transmission?

A. Listening for novice stations

B. Operating full break-in

C. Listening only for a specific station or stations

D. Closing station now

Plain-English Study Guide for the FCC Amateur Radio General Class License

What prosign is sent to indicate the end of a formal message when using CW?

A. SK

B. BK

C. AR

D. KN

What does the Q signal "QSL" mean?

A. Send slower

B. We have already confirmed by card

C. I acknowledge receipt

D. We have worked before

What does the Q signal "QRV" mean?

A. You are sending too fast

B. There is interference on the frequency

C. I am quitting for the day

D. I am ready to receive messages

What does the Q signal "QRN" mean?

A. Send more slowly

B. Stop sending

C. Zero beat my signal

D. *I am troubled by static*

What is QRP operation?

A. Remote piloted model control

B. *Low power transmit operation*

C. Transmission using Quick Response Protocol

D. Traffic relay procedure net operation

What is meant by the term MMIC?

A. Multi-Megabyte Integrated Circuit

B. *Monolithic Microwave Integrated Circuit*

C. Military Manufactured Integrated Circuit

D. Mode Modulated Integrated Circuit

What is meant by the term ROM?

A. Resistor Operated Memory

B. *Read Only Memory*

C. Random Operational Memory

D. Resistant to Overload Memory

Plain-English Study Guide for the FCC Amateur Radio General Class License

11 EQUIPMENT AND ITS OPERATION

The test covers the operation of common controls on your receiver, transmitter, or transceiver.

You need to know the names of some parts of a **superheterodyne** receiver. **Heterodyning** means the mixing of two signals, and a superheterodyne receiver works by mixing the incoming signal with a signal from the receiver's own **local oscillator** (sometimes called the **HF oscillator** on the test). To tune the receiver, the frequency of the local oscillator is what changes. The part of the receiver that combines the two signals is called the **mixer**. It combines the incoming signal with the local oscillator and outputs the **intermediate frequency (IF)**, which is the difference between the two inputs, (The other output is the sum, but this is usually not used. But there is a question on the test where you have to know that the output is the sum and difference.) For example, in one question on the test, the input signal is 14.250 MHz, the local oscillator is 13.795 Mhz, and the IF is 0.455 MHz, or 455 kHz.

One type of interference that is covered on the test is **image response**. This is caused by the mixer "seeing" an input frequency which is two times the IF lower (or higher) than the desired frequency. For example, on the test, there is a question about listening to a signal on 14.255 MHz on a receiver with an IF of 455 kHz. It is receiving interference from a signal on 13.345 MHz, which is 910 kHz (2 times 455) lower than the desired signal. This is probably an image response.

The simplest possible superheterodyne receiver would contain an HF oscillator, a mixer, and a detector. The **mixer** is the part that processes signals from the RF amplifier and local oscillator and sends them to the IF filter. In an SSB receiver, the product detector is the part that combines the signals from the IF amplifier and BFO and sends them to the AF amplifier. A **notch filter** is used to reduce interference from a carrier in the receiver passband by "notching it out".

The **discriminator** is the part of an FM receiver that converts signals coming from the IF amplifier to audio.

The **IF shift** control on a receiver also can be used to avoid interference from stations very close to the receive frequency.

Many receivers have an **attenuator** to reduce signal overload due to strong incoming signals.

Many receivers also have a control called a **noise blanker**, which works by turning down the receiver's gain during a pulse of noise. If this control is turned up too high, it may cause received signals to become distorted.

Some receivers have a **digital signal processor (DSP)** to remove noise from received signals. The advantage of a DSP filter over an analog filter is that a wide range of filter bandwidths and shapes can be created. A DSP filter can also be set to automatically "notch out" an interfering carrier. DSP filtering is accomplished by converting the signal from analog to digital and using digital processing. Therefore, a DSP filter contains all of the following: an analog to digital converter, a digital to analog converter, and a digital processor chip.

A **software-defined radio (SDR)** is a radio in which most major signal processing functions are performed by software. For the test, you need to know the following two SDR facts: I and Q signals are used **to create all**

Plain-English Study Guide for the FCC Amateur Radio General Class License

types of modulation. The phase difference between the two is **90 degrees**.

Many receivers have adjustable bandwidth. It's a good idea to match the receiver bandwidth to the bandwidth of the transmitted signal, because this yields the best signal-to-noise ratio.

If you are using your transceiver in **split** mode, this means that you are transmitting and receiving on different frequencies. Many transceivers have a **dual VFO** to make it easier to monitor the transmit and receive frequencies when they are not the same.

Some transmitters have a **speech processor**. Its purpose is to increase the intelligibility of transmitted phone signals during poor conditions. It does this by increasing the average power. But if incorrectly adjusted, it can cause distorted speech, splatter, and excessive background pickup.

Many transmitters have a time delay in the keying circuit. This is to allow time for transmit-receive changeover operations to complete properly before RF output is allowed.

Some transceivers are controlled by a **direct digital synthesizer (DDS)**. This has the advantage of giving variable frequency control with the stability of a crystal oscillator.

You need to know the definitions of the following types of modulation. Amplitude **modulation (AM)** is the process that changes the envelope of an RF wave to carry information. Another way of saying this is that AM varies the instantaneous power level. One question asks you what the modulation envelope is. This is just the name given to the waveform that is created.

Phase modulation (PM) is the process that changes the phase angle to convey information. A **reactance modulator** is used to produce phase modulation. **Frequency modulation (FM)** is the process which changes the frequency to convey information. One question uses more confusing language that says the same thing: When an audio signal is applied to an FM transmitter, the carrier frequency changes proportionally to the instantaneous amplitude of the modulating signal.

The test covers a number of questions about amplifiers. Different amplifiers can be of different "classes". The classes on the test are **Class A** and **Class C**. (The other two classes, B and AB are mentioned on the test, but only as wrong answers). For the test, you need to know that a Class A amplifier has low distortion. A Class C amplifier has the highest efficiency, but can only be used for FM (or CW, but that's not on the test). An amplifier is **linear** if the output preserves the input waveform.

The final amplifier stage of a transmitter must be **neutralized**. (This involves adjusting a neutralizing capacitor.) This must be done to eliminate self-oscillation.

The efficiency of an amplifier is the RF output power divided by the DC input power. The **ALC (Automatic Level Control)** of a transmitter or amplifier is used to reduce distortion due to excessive drive. (Normally, the ALC is adjusted by adjusting the transmit audio or microphone gain.)

The test will have some questions about how to tune a tube amplifier. To tune such an amplifier, there are generally two controls. The first control is the "Plate Tuning" control. This is adjusted so that the plate current meter shows a pronounced dip. The other control is the "Load" or "Coupling" control. This is set to the maximum power output (without exceeding the maximum allowable plate current).

In the case of a solid state amplifier, it is important not to use excessive drive power, because this can cause permanent damage.

There are a few questions on the test about **oscillators**, which are used in both transmitters and receivers. Virtually all sine wave oscillators contain a filter and an amplifier operating in a feedback loop. In the case of an LC oscillator, the frequency is determined by the inductance and capacitance in the tank circuit.

Plain-English Study Guide for the FCC Amateur Radio General Class License

What is the modulation envelope of an AM signal?

A. The waveform created by connecting the peak values of the modulated signal

B. The carrier frequency that contains the signal

C. Spurious signals that envelop nearby frequencies

D. The bandwidth of the modulated signal

What is the purpose of the "notch filter" found on many HF transceivers?

A. To restrict the transmitter voice bandwidth

B. To reduce interference from carriers in the receiver passband

C. To eliminate receiver interference from impulse noise sources

D. To enhance the reception of a specific frequency on a crowded band

What is normally meant by operating a transceiver in "split" mode?

A. The radio is operating at half power

B. The transceiver is operating from an external power source

C. The transceiver is set to different transmit and receive frequencies

D. The transmitter is emitting a SSB signal, as opposed to DSB operation

Which of the following is a use for the IF shift control on a receiver?

A. To avoid interference from stations very close to the receive frequency

B. To change frequency rapidly

C. To permit listening on a different frequency from that on which you are transmitting

D. To tune in stations that are slightly off frequency without changing your transmit frequency

Which of the following is a common use for the dual VFO feature on a transceiver?

A. To allow transmitting on two frequencies at once

B. To permit full duplex operation, that is transmitting and receiving at the same time

C. To permit monitoring of two different frequencies

D. To facilitate computer interface

What is one reason to use the attenuator function that is present on many HF transceivers?

A. To reduce signal overload due to strong incoming signals

B. To reduce the transmitter power when driving a linear amplifier

C. To reduce power consumption when operating from batteries

D. To slow down received CW signals for better copy

What reading on the plate current meter of a vacuum tube RF power amplifier indicates correct adjustment of the plate tuning control?

A. A pronounced peak

B. A pronounced dip

C. No change will be observed

D. A slow, rhythmic oscillation

What is the correct adjustment for the load or coupling control of a vacuum tube RF power amplifier?

A. Minimum SWR on the antenna

B. Minimum plate current without exceeding maximum allowable grid current

C. Highest plate voltage while minimizing grid current

D. Maximum power output without exceeding maximum allowable plate current

What condition can lead to permanent damage when using a solid-state RF power amplifier?

A. Exceeding the Maximum Usable Frequency

B. Low input SWR

C. Shorting the input signal to ground

D. Excessive drive power

What is a reason to use Automatic Level Control (ALC) with an RF power amplifier?

A. To balance the transmitter audio frequency response

B. To reduce harmonic radiation

C. To reduce distortion due to excessive drive

D. To increase overall efficiency

Why is a time delay sometimes included in a transmitter keying circuit?

A. To prevent stations from talking over each other

B. To allow the transmitter power regulators to charge properly

C. To allow time for transmit-receive changeover operations to complete properly before RF output is allowed

D. To allow time for a warning signal to be sent to other stations

How does a noise blanker work?

A. By temporarily increasing received bandwidth

B. By redirecting noise pulses into a filter capacitor

C. By reducing receiver gain during a noise pulse

D. By clipping noise peaks

Plain-English Study Guide for the FCC Amateur Radio General Class License

What happens as the noise reduction control level in a receiver is increased?

A. Received signals may become distorted

B. Received frequency may become unstable

C. CW signals may become severely attenuated

D. Received frequency may shift several kHz

What is the phase difference between the I and Q signals that software-defined radio (SDR) equipment uses for modulation and demodulation?

A. Zero

B. 90 degrees

C. 180 degrees

D. 45 degrees

What is an advantage of using I and Q signals in software-defined radios (SDRs)?

A. The need for high resolution analog-to-digital converters is eliminated

B. All types of modulation can be created with appropriate processing

C. Minimum detectible signal level is reduced

D. Converting the signal from digital to analog creates mixing products

Which of the following is an advantage of a receiver DSP IF filter as compared to an analog filter?

A. A wide range of filter bandwidths and shapes can be created

B. Fewer digital components are required

C. Mixing products are greatly reduced

D. The DSP filter is much more effective at VHF frequencies

What is the purpose of a speech processor as used in a modern transceiver?

A. Increase the intelligibility of transmitted phone signals during poor conditions

B. Increase transmitter bass response for more natural sounding SSB signals

C. Prevent distortion of voice signals

D. Decrease high-frequency voice output to prevent out of band operation

Which of the following describes how a speech processor affects a transmitted single sideband phone signal?

A. It increases peak power

B. It increases average power

C. It reduces harmonic distortion

D. It reduces intermodulation distortion

Which of the following can be the result of an incorrectly adjusted speech processor?

A. Distorted speech

B. Splatter

C. Excessive background pickup

D. All of these choices are correct

Which of the following are basic components of a sine wave oscillator?

A. An amplifier and a divider

B. A frequency multiplier and a mixer

C. A circulator and a filter operating in a feed-forward loop

D. A filter and an amplifier operating in a feedback loop

How is the efficiency of an RF power amplifier determined?

A. Divide the DC input power by the DC output power

B. Divide the RF output power by the DC input power

C. Multiply the RF input power by the reciprocal of the RF output power

D. Add the RF input power to the DC output power

What determines the frequency of an LC oscillator?

A. The number of stages in the counter

B. The number of stages in the divider

C. The inductance and capacitance in the tank circuit

D. The time delay of the lag circuit

For which of the following modes is a Class C power stage appropriate for amplifying a modulated signal?

A. SSB

B. FM

C. AM

D. All these choices are correct

Which of these classes of amplifiers has the highest efficiency?

A. Class A

B. Class B

C. Class AB

D. Class C

What is the reason for neutralizing the final amplifier stage of a transmitter?

A. To limit the modulation index

B. To eliminate self-oscillations

C. To cut off the final amplifier during standby periods

D. To keep the carrier on frequency

Which of the following describes a linear amplifier?

A. Any RF power amplifier used in conjunction with an amateur transceiver

B. An amplifier in which the output preserves the input waveform

C. A Class C high efficiency amplifier

D. An amplifier used as a frequency multiplier

What circuit is used to process signals from the RF amplifier and local oscillator then send the result to the IF filter in a superheterodyne receiver?

A. Balanced modulator

B. IF amplifier

C. Mixer

D. Detector

What combination of a mixer's Local Oscillator (LO) and RF input frequencies is found in the output?

A. The ratio

B. The average

C. The sum and difference

D. The arithmetic product

Which mixer input is varied or tuned to convert signals of different frequencies to an intermediate frequency (IF)?

A. Image frequency

B. Local oscillator

C. RF input

D. Beat frequency oscillator

What circuit is used to combine signals from the IF amplifier and BFO and send the result to the AF amplifier in some single sideband receivers?

A. RF oscillator

B. IF filter

C. Balanced modulator

D. Product detector

Plain-English Study Guide for the FCC Amateur Radio General Class License

Which of the following is an advantage of a transceiver controlled by a direct digital synthesizer (DDS)?

A. Wide tuning range and no need for band switching

B. Relatively high power output

C. Relatively low power consumption

D. Variable frequency with the stability of a crystal oscillator

Which of the following is a typical application for a Direct Digital Synthesizer?

A. A high-stability variable frequency oscillator in a transceiver

B. A digital voltmeter

C. A digital mode interface between a computer and a transceiver

D. A high-sensitivity radio direction finder

What is the simplest combination of stages that implement a superheterodyne receiver?

A. RF amplifier, detector, audio amplifier

B. RF amplifier, mixer, IF discriminator

C. HF oscillator, mixer, detector

D. HF oscillator, pre-scaler, audio amplifier

What type of circuit is used in many FM receivers to convert signals coming from the IF amplifier to audio?

A. Product detector

B. Phase inverter

C. Mixer

D. Discriminator

What is meant by the term "software defined radio" (SDR)?

A. *A radio in which most major signal processing functions are performed by software*

B. A radio which provides computer interface for automatic logging of band and frequency

C. A radio which uses crystal filters designed using software

D. A computer model which can simulate performance of a radio to aid in the design process

What is the name of the process that changes the phase angle of an RF wave to convey information?

A. Phase convolution

B. *Phase modulation*

C. Angle convolution

D. Radian inversion

What is the name of the process that changes the instantaneous frequency of an RF wave to convey information?

A. Frequency convolution

B. Frequency transformation

C. Frequency conversion

D. Frequency modulation

What emission is produced by a reactance modulator connected to an RF power amplifier?

A. Multiplex modulation

B. Phase modulation

C. Amplitude modulation

D. Pulse modulation

What type of modulation varies the instantaneous power level of the RF signal?

A. Frequency shift keying

B. Pulse position modulation

C. Frequency modulation

D. Amplitude modulation

What control is typically adjusted for proper ALC setting on an amateur single sideband transceiver?

A. The RF clipping level

B. Transmit audio or microphone gain

C. Antenna inductance or capacitance

D. Attenuator level

If a receiver mixes a 13.800 MHz VFO with a 14.255 MHz received signal to produce a 455 kHz intermediate frequency (IF) signal, what type of interference will a 13.345 MHz signal produce in the receiver?

A. Quadrature noise

B. Image response

C. Mixer interference

D. Intermediate interference

What is another term for the mixing of two RF signals?

A. Heterodyning

B. Synthesizing

C. Cancellation

D. Phase inverting

Why is it good to match receiver bandwidth to the bandwidth of the operating mode?

A. It is required by FCC rules

B. It minimizes power consumption in the receiver

C. It improves impedance matching of the antenna

D. It results in the best signal to noise ratio

Richard P. Clem

12 SINGLE SIDEBAND

The most popular phone (voice) mode on the HF bands is **Single Sideband (SSB)**, which can be either Lower Sideband (LSB) or Upper Sideband (USB). SSB is used because it uses less bandwidth than other modes, and it has a higher power efficiency. When you transmit an SSB signal, only one sideband is being transmitted. The other two parts of the signal--the carrier and the other sideband--are **suppressed**, meaning that they are not transmitted. The carrier is suppressed so that the available transmitter power can be used more effectively. There is one question that asks which of four common phone mode uses the least bandwidth. (The choices are SSB, DSB, PM, and FM.) The correct answer is SSB.

The typical SSB signal has a bandwidth of just under 3 kHz. Therefore, the customary minimum frequency separation between SSB signals is approximately 3 kHz.

It is important to understand exactly what frequencies your transmitted signal covers, because the entire signal must stay within the authorized frequency range. You are not transmitting on just one frequency—you are really transmitting on a band of frequencies about 3 kHz wide. In other words, the frequency shown on your transmitter is not the only frequency on which you are transmitting. In fact, the dial of an SSB transmitter usually displays the carrier frequency. And as we have learned, the carrier is actually suppressed, meaning that it is not being transmitted. So the frequency on the dial is actually one frequency on which you are not transmitting!

If you are using Upper Sideband (USB), then you are really transmitting just above the frequency shown on your dial. And if you are using Lower Sideband (LSB), then you are really transmitting just below the frequency shown on your dial. The width of most SSB signals is about 3 kHz. For example, if you are transmitting LSB and your transmitter's dial reads 7178 kHz, then your signal is actually being transmitted on all frequencies between 7175 and 7178 kHz. Or if you are using USB on 14347 kHz, then your signal is actually being transmitted on all frequencies between 14347 and 14350 kHz.

Therefore, if you are using SSB, you need to stay at least 3 kHz away from the band edge: If using LSB, you need to stay 3 kHz away from the lower band edge. If using USB, you need to stay 3 kHz away from the upper band edge. For example, you're not allowed above 14350. Therefore, you can't set your dial any higher than 14347 when using USB. And if you're not allowed below 7175, then you can't set your dial any lower than 7178 when using LSB. Of course, it is wise to use a slightly larger "cushion" in case your dial is not completely accurate. But when using LSB, you can never go closer than 3 kHz away from the lower band edge. And when using USB, you can never go closer than 3 kHz away from the upper band edge.

Upper Sideband is most commonly used on the frequencies 14 MHz and above, and Lower Sideband is used on the frequencies below 14 MHz. (Remember, lower sideband is used on the lower frequencies, and upper sideband is used on the upper frequencies.) This is not a legal requirement; it is just the current amateur practice.

To break in to an existing conversation on any voice mode, the recommended method is to simply say your callsign during a break between transmissions.

VOX stands for "voice operated switch", and is used for "hands free" operation.

There are two parts of an SSB transmitter whose names you must know for the test. The **balanced modulator** is the part that combines signals from the carrier oscillator and speech amplifier. The output of the balance modulator is both the upper and lower sidebands, one of which is later filtered out. The **filter** is the part of the transmitter that processes signals from the balanced modulator and sends them to the mixer.

Over-modulation can cause excessive bandwidth. Also, excessive drive of an SSB transmitter can cause a type of distortion called **flat-topping**.

Which sideband is most commonly used for voice communications on frequencies of 14 MHz or higher?

A. Upper sideband

B. Lower sideband

C. Vestigial sideband

D. Double sideband

Which of the following modes is most commonly used for voice communications on the 160, 75, and 40 meter bands?

A. Upper sideband

B. Lower sideband

C. Vestigial sideband

D. Double sideband

Plain-English Study Guide for the FCC Amateur Radio General Class License

Which of the following is most commonly used for SSB voice communications in the VHF and UHF bands?

A. Upper sideband

B. Lower sideband

C. Vestigial sideband

D. Double sideband

Which mode is most commonly used for voice communications on the 17 and 12 meter bands?

A. Upper sideband

B. Lower sideband

C. Vestigial sideband

D. Double sideband

Which mode of voice communication is most commonly used on the HF amateur bands?

A. Frequency modulation

B. Double sideband

C. Single sideband

D. Phase modulation

G2A06 (D)

Which of the following is an advantage when using single sideband, as compared to other analog voice modes on the HF amateur bands?

A. Very high fidelity voice modulation

B. Less subject to interference from atmospheric static crashes

C. Ease of tuning on receive and immunity to impulse noise

D. Less bandwidth used and greater power efficiency

Which of the following statements is true of the single sideband (SSB) voice mode?

A. Only one sideband and the carrier are transmitted; the other sideband is suppressed

B. Only one sideband is transmitted; the other sideband and carrier are suppressed

C. SSB is the only voice mode that is authorized on the 20-meter, 15-meter, and 10-meter amateur bands

D. SSB is the only mode that is authorized on the 160, 75 and 40 meter amateur bands

What is the recommended way to break in to a phone contact?

A. Say "QRZ" several times, followed by your call sign

B. Say your call sign once

C. Say "Breaker Breaker"

D. Say "CQ" followed by the call sign of either station

Why do most amateur stations use lower sideband on the 160-meter, 75-meter, and 40-meter bands?

A. Lower sideband is more efficient than upper sideband at these frequencies

B. Lower sideband is the only sideband legal on these frequency bands

C. Because it is fully compatible with an AM detector

D. It is good amateur practice

Which of the following statements is true of voice VOX operation versus PTT operation?

A. The received signal is more natural sounding

B. It allows "hands free" operation

C. It occupies less bandwidth

D. It provides more power output

What is the customary minimum frequency separation between SSB signals under normal conditions?

A. Between 150 and 500 Hz

B. Approximately 3 kHz

C. Approximately 6 kHz

D. Approximately 10 kHz

What frequency range is occupied by a 3 kHz LSB signal when the displayed carrier frequency is set to 7.178 MHz?

A. 7.178 to 7.181 MHz

B. 7.178 to 7.184 MHz

C. 7.175 to 7.178 MHz

D. 7.1765 to 7.1795 MHz

What frequency range is occupied by a 3 kHz USB signal with the displayed carrier frequency set to 14.347 MHz?

A. 14.347 to 14.647 MHz

B. 14.347 to 14.350 MHz

C. 14.344 to 14.347 MHz

D. 14.3455 to 14.3485 MHz

How close to the lower edge of the 40-meter General Class phone segment should your displayed carrier frequency be when using 3 kHz wide LSB?

A. At least 3 kHz above the edge of the segment

B. At least 3 kHz below the edge of the segment

C. Your displayed carrier frequency may be set at the edge of the segment

D. At least 1 kHz above the edge of the segment

How close to the upper edge of the 20-meter General Class band should your displayed carrier frequency be when using 3 kHz wide USB?

A. At least 3 kHz above the edge of the band

B. At least 3 kHz below the edge of the band

C. Your displayed carrier frequency may be set at the edge of the band

D. At least 1 kHz below the edge of the segment

Which of the following is used to process signals from the balanced modulator and send them to the mixer in a single-sideband phone transmitter?

A. Carrier oscillator

B. Filter

C. IF amplifier

D. RF amplifier

Which circuit is used to combine signals from the carrier oscillator and speech amplifier and send the result to the filter in a typical single-sideband phone transmitter?

A. Discriminator

B. Detector

C. IF amplifier

D. Balanced modulator

Which of the following phone emissions uses the narrowest frequency bandwidth?

A. Single sideband

B. Double sideband

C. Phase modulation

D. Frequency modulation

Which of the following is an effect of overmodulation?

A. Insufficient audio

B. Insufficient bandwidth

C. Frequency drift

D. Excessive bandwidth

What is meant by the term "flat-topping," when referring to a single sideband phone transmission?

A. Signal distortion caused by insufficient collector current

B. The transmitter's automatic level control (ALC) is properly adjusted

C. *Signal distortion caused by excessive drive*

D. The transmitter's carrier is properly suppressed

Richard P. Clem

13 FREQUENCY MODULATION

There are a few more questions about FM that we haven't covered in other chapters. Most FM operation takes place above 29.5 MHz. This is because most FM signals have a wide bandwidth that is not allowed below that frequency. Because it is easier to build oscillators for lower frequencies, most VHF transmitters operate by generating a low frequency and then using a multiplier stage to reach the desired operating frequency.

There are two questions on the test about the bandwidth and deviation of an FM signal. It is probably easiest to simply memorize the answer to these questions. In one question, the FM signal has 5 kHz deviation and a 2 kHz modulating frequency. The bandwidth of this signal is 16 kHz.

The other question asks for the deviation of a 146.52 MHz signal that was modulated at a lower frequency. The correct answer to this question is 416.7 Hz.

G8B04 (D)

What is the stage in a VHF FM transmitter that generates a harmonic of a lower frequency signal to reach the desired operating frequency?

A. Mixer

B. Reactance modulator

C. Pre-emphasis network

D. Multiplier

What is the total bandwidth of an FM-phone transmission having a 5 kHz deviation and a 3 kHz modulating frequency?

A. 3 kHz

B. 5 kHz

C. 8 kHz

D. 16 kHz

What is the frequency deviation for a 12.21-MHz reactance-modulated oscillator in a 5-kHz deviation, 146.52-MHz FM-phone transmitter?

A. 101.75 Hz

B. 416.7 Hz

C. 5 kHz

D. 60 kHz

Richard P. Clem

14 DIGITAL MODES

There are many digital modes available, and you need to know a little bit about many of them for the test. In any kind of digital communication, higher symbol rates require higher bandwidths.

One of the oldest digital modes is **RTTY (radioteletype)** using **AFSK (audio frequency shift keying)**. The most common frequency shift used on the amateur bands for RTTY is 170 Hz. It is called frequency shift keying because the frequency shifts between two frequencies, called the "mark" and the "space." In AFSK, the digital signal is converted into an audio signal, and this audio signal is used to modulate an SSB transmitter. Normally, LSB is used for RTTY. However, for other digital modes, such as JT9, JT65, or FT8, USB is normally used. (JT9, JT65, and FT8 are modes that especially good for use with extremely low signal strength.)

RTTY commonly uses the **Baudot** code, which is a 5-bit code with additional start and stop bits.

One way to generate an FSK signal is with AFSK, using an audio signal. The other way is to change an oscillator's frequency directly with a digital control signal. If you can't decode a RTTY or other FSK signal, here are some of the things that might be wrong: What the mark and space frequencies may be reversed, you may have the wrong baud rate, or you may be listening on the wrong sideband.

PSK31 is a popular "sound card" mode, since it is a simple matter to hook a transceiver to a computer's sound card, and the PSK31 mode is very efficient. The number 31 in PSK31 refers to the approximate transmitted symbol rate. When transmitting PSK31, the transmitter audio input should be adjusted so that the transceiver ALC system does not activate. The number of data bits in a single PSK31 character varies. For that reason, it is called a **varicode.** Upper case letters have longer symbols than lower case letters. Therefore, upper case letters take longer to send and slow down the transmission. (For the test, you also need to know that QPSK31 and BPSK31 both have about the same bandwidth, they are both sideband sensitive, and the encoding has error correction.)

PSK31 generally uses a "waterfall" display on the computer. It is a graph on the computer screen where signals flow down from the top like a waterfall. The horizontal position shows frequency, and the vertical position shows time. The strength of the signal is shown by the intensity on the screen. If there are vertical lines next to a station's signal on the waterfall, this generally indicates overmodulation.

With any soundcard mode, it is important to set the modulation level and ALC correctly. If it is set too high, this will cause overmodulation, which will result in distortion and spurious emissions.

You also need to avoid having audio cables pick up RF interference. If they do, this can result in problems such as the VOX circuit not un-keying the transmitter, distortion, or frequent connection timeouts.

FT8 is the name of a relatively new digital mode, and there are some questions about it on the test. It can be used even when signals are extremely weak—in other words, when the signal-to-noise ratio is very low. You need to know that it uses 8-tone frequency shift keying and that the amount of information you can send is very limited. It is generally limited to call signs, grid squares, and signal reports. You also need to know that to use FT8, your computer's clock must be accurate to within about one second.

There is one question about a digital mode named **WSPR**. You need to know that WSPR is used as a low-power beacon to assess HF propagation.

Some digital modes use error correction. **Forward error correction** allows the receiver to correct errors by transmitting redundant information with the data.

Packet radio is another means of digital communication. Each data packet sent contains a **header**, which contains the routing and handling information for the message. Packet radio employs error correction. If the receiving station receives a packet containing errors, it requests that the packet be retransmitted.

In the **PACTOR** protocol, an **NAK** response means that the receiver is requesting that the packet be re-transmitted. PACTOR is a mode used only for communications between two stations. There is a trick question on the test about how to break in to a PACTOR contact. This is impossible, because you can't break in to the contact.

At maximum data rate, a PACTOR signal has a bandwidth of about 2300 Hz (which is about the same as an SSB voice signal). In the PACTOR and WINMOR modes, if there is a failure to exchange information after too many attempts, then the connection is dropped.

When using any mode, you should always check to see if the frequency is in use before transmitting. This is especially important with digital modes, since someone using a different mode on the same frequency might not show up on your computer. With PACTOR, the easiest way to do this is to put the modem or controller in a mode which allows monitoring communications without a connection

PACTOR or WINMOR interference can result in frequent retries or timeouts, long pauses in message transmission, and failure to establish a connection between stations.

Plain-English Study Guide for the FCC Amateur Radio General Class License

There are a few questions on the test about digital messaging system gateway stations. One such system is Winlink, a system which uses the Internet to transfer messages. To contact such a station, you transmit the correct message on that station's published frequency.

There are also a few miscellaneous digital questions on the test. You need to know the following facts: One advantage of using the binary system when processing digital signals is because binary "ones" and "zeros" are easy to represent with an "on" or "off" state. A three bit binary counter has 8 possible states (2 times 2 times 2). A shift register is a clocked array of circuits that passes data in steps along the array.

There is one trick question on the test that asks when digital modes are exempt from Part 97 of the FCC rules. This is a trick question because they are never exempt from the rules.

Which mode is normally used when sending an RTTY signal via AFSK with an SSB transmitter?

A. USB

B. DSB

C. CW

D. LSB

What is the standard sideband used to generate a JT65, JT9, or FT8 digital signal when using AFSK in any amateur band?

A. LSB

B. USB

145

C. DSB

D. SSB

What part of a data packet contains the routing and handling information?

A. Directory

B. Preamble

C. Header

D. Footer

Which of the following describes Baudot code?

A. A 7-bit code with start, stop and parity bits

B. A code using error detection and correction

C. A 5-bit code with additional start and stop bits

D. A code using SELCAL and LISTEN

What is the most common frequency shift for RTTY emissions in the amateur HF bands?

A. 85 Hz

B. 170 Hz

C. 425 Hz

D. 850 Hz

How does the receiving station respond to an ARQ data mode packet containing errors?

A. Terminates the contact

B. Requests the packet be retransmitted

C. Sends the packet back to the transmitting station

D. Requests a change in transmitting protocol

In the PACTOR protocol, what is meant by an NAK response to a transmitted packet?

A. The receiver is requesting the packet be re-transmitted

B. The receiver is reporting the packet was received without error

C. The receiver is busy decoding the packet

D. The entire file has been received correctly

How many states does a 3-bit binary counter have?

A. 3

B. 6

C. 8

D. 16

What is a shift register?

A. A clocked array of circuits that passes data in steps along the array

B. An array of operational amplifiers used for tri state arithmetic operations

C. A digital mixer

D. An analog mixer

What does the number 31 represent in PSK31?

A. The approximate transmitted symbol rate

B. The version of the PSK protocol

C. The year in which PSK31 was invented

D. The number of characters that can be represented by PSK31

Which of the following is characteristic of QPSK31?

A. It is sideband sensitive

B. Its encoding provides error correction

C. Its bandwidth is approximately the same as BPSK31

D. All these choices are correct

How does forward error correction (FEC) allow the receiver to correct errors in received data packets?

A. By controlling transmitter output power for optimum signal strength

B. By using the varicode character set

C. By transmitting redundant information with the data

D. By using a parity bit with each character

What is the relationship between transmitted symbol rate and bandwidth?

A. Symbol rate and bandwidth are not related

B. Higher symbol rates require wider bandwidth

C. Lower symbol rates require wider bandwidth

D. Bandwidth is always half the symbol rate

Which of the following is characteristic of the FT8 mode of the WSJT-X family?

A. It is a keyboard-to-keyboard chat mode

B. Each transmission takes exactly 60 seconds

C. It is limited to use on VHF

D. Typical exchanges are limited to call signs, grid locators, and signal reports

Which of the following is a requirement when using the FT8 digital mode?

A. A special hardware modem

B. Computer time accurate within approximately 1 second

C. Receiver attenuator set to -12 dB

D. A vertically polarized antenna

What type of modulation is used by the FT8 digital mode?

A. 8-tone frequency shift keying

B. Vestigial sideband

C. Amplitude compressed AM

D. Direct sequence spread spectrum

Which of the following narrow-band digital modes can receive signals with very low signal-to-noise ratios?

A. MSK144

B. FT8

C. AMTOR

D. MFSK32

Which digital mode is used as a low-power beacon for assessing HF propagation?

A. WSPR

B. Olivia

C. PSK31

D. SSB-SC

Plain-English Study Guide for the FCC Amateur Radio General Class License

How can a PACTOR modem or controller be used to determine if the channel is in use by other PACTOR stations?

A. Unplug the data connector temporarily and see if the channel-busy indication is turned off

B. Put the modem or controller in a mode which allows monitoring communications without a connection

C. Transmit UI packets several times and wait to see if there is a response from another PACTOR station

D. Send the message: "Is this frequency in use?"

What symptoms may result from other signals interfering with a PACTOR or WINMOR transmission?

A. Frequent retries or timeouts

B. Long pauses in message transmission

C. Failure to establish a connection between stations

D. All of these choices are correct

How do you join a contact between two stations using the PACTOR protocol?

A. Send broadcast packets containing your call sign while in MONITOR mode

B. Transmit a steady carrier until the PACTOR protocol times out and disconnects

C. Joining an existing contact is not possible, PACTOR connections are limited to two stations

D. Send a NAK response continuously so that the sending station has to pause

Which of the following is a way to establish contact with a digital messaging system gateway station?

A. Send an email to the system control operator

B. Send QRL in Morse code

C. Respond when the station broadcasts its SSID

D. Transmit a connect message on the station's published frequency

Which communication system sometimes uses the internet to transfer messages?

A. Winlink

B. RTTY

C. ARES

D. SKYWARN

What is indicated on a waterfall display by one or more vertical lines adjacent to a PSK31 signal?

A. Long Path propagation

B. Backscatter propagation

C. Insufficient modulation

D. Overmodulation

Which of the following describes a waterfall display?

A. Frequency is horizontal, signal strength is vertical, time is intensity

B. Frequency is vertical, signal strength is intensity, time is horizontal

C. Frequency is horizontal, signal strength is intensity, time is vertical

D. Frequency is vertical, signal strength is horizontal, time is intensity

What could be wrong if you cannot decode an RTTY or other FSK signal even though it is apparently tuned in properly?

A. The mark and space frequencies may be reversed

B. You may have selected the wrong baud rate

C. You may be listening on the wrong sideband

D. All of these choices are correct

What is likely to happen if a transceiver's ALC system is not set properly when transmitting AFSK signals with the radio using single sideband mode?

A. ALC will invert the modulation of the AFSK mode

B. Improper action of ALC distorts the signal and can cause spurious emissions

C. When using digital modes, too much ALC activity can cause the transmitter to overheat

D. All of these choices are correct

Plain-English Study Guide for the FCC Amateur Radio General Class License

Which of the following can be a symptom of transmitted RF being picked up by an audio cable carrying AFSK data signals between a computer and a transceiver?

A. The VOX circuit does not un-key the transmitter

B. The transmitter signal is distorted

C. Frequent connection timeouts

D. All of these choices are correct

How is an FSK signal generated?

A. By keying an FM transmitter with a sub-audible tone

B. By changing an oscillator's frequency directly with a digital control signal

C. By using a transceiver's computer data interface protocol to change frequencies

D. By reconfiguring the CW keying input to act as a tone generator

What is the approximate bandwidth of a PACTOR-III signal at maximum data rate?

A. 31.5 Hz

B. 500 Hz

C. 1800 Hz

D. 2300 Hz

What action results from a failure to exchange information due to excessive transmission attempts when using PACTOR or WINMOR?

A. The checksum overflows

B. *The connection is dropped*

C. Packets will be routed incorrectly

D. Encoding reverts to the default character set

Which of the following statements is true about PSK31?

A. Upper case letters make the signal stronger

B. *Upper case letters use longer Varicode signals and thus slow down transmission*

C. Varicode Error Correction is used to ensure accurate message reception

D. Higher power is needed as compared to RTTY for similar error rates

How are the two separate frequencies of a Frequency Shift Keyed (FSK) signal identified?

A. Dot and Dash

B. On and Off

C. High and Low

D. Mark and Space

Which type of code is used for sending characters in a PSK31 signal?

A. Varicode

B. Viterbi

C. Volumetric

D. Binary

Under what circumstances are messages that are sent via digital modes exempt from Part 97 third party rules that apply to other modes of communication?

A. Under no circumstances

B. When messages are encrypted

C. When messages are not encrypted

D. When under automatic control

Richard P. Clem

15 MOBILE OPERATION

The test contains a few questions about setting up a mobile station.

It is best to hook up the power connections for a mobile radio directly to the car's battery, using heavy gauge wire. It's not a good idea to use the car's auxiliary power socket (formerly known as the cigarette lighter socket) because the socket's wiring might be inadequate for the radio's current draw.

There are two questions about things that are commonly found on top of mobile antennas. A capacitance hat is a device that electrically lengthens a physically short antenna. And a corona ball is a small round ball that reduces high voltage discharge from the tip of the antenna (and to lessen the damage from things getting poked by the antenna, but according to the test, that's a wrong answer).

One question asks what the main limiting factor is for a 75 meter mobile radio. The answer is the antenna system. Normally, an antenna for 75 meters needs to be very long, and it becomes very inefficient if it's shortened enough to fit on a car. In addition, when an antenna is shortened, the bandwidth of the antenna is usually very limited.

The car's vehicle control computer can often radiate signals that are picked up by your receiver, and this is the most common source of interfering signals in modern cars.

What is the purpose of a capacitance hat on a mobile antenna?

A. To increase the power handling capacity of a whip antenna

B. To allow automatic band changing

C. To electrically lengthen a physically short antenna

D. To allow remote tuning

What is the purpose of a corona ball on an HF mobile antenna?

A. To narrow the operating bandwidth of the antenna

B. To increase the "Q" of the antenna

C. To reduce the chance of damage if the antenna should strike an object

D. To reduce RF voltage discharge from the tip of the antenna while transmitting

Which of the following direct, fused power connections would be the best for a 100-watt HF mobile installation?

A. To the battery using heavy gauge wire

B. To the alternator or generator using heavy gauge wire

C. To the battery using resistor wire

D. To the alternator or generator using resistor wire

Richard P. Clem

Why is it best NOT to draw the DC power for a 100-watt HF transceiver from an automobile's auxiliary power socket?

A. The socket is not wired with an RF-shielded power cable

B. The socket's wiring may be inadequate for the current being drawn by the transceiver

C. The DC polarity of the socket is reversed from the polarity of modern HF

transceivers

D. Drawing more than 50 watts from this socket could cause the engine to overheat

Which of the following most limits the effectiveness of an HF mobile transceiver operating in the 75 meter band?

A. "Picket Fencing" signal variation

B. The wire gauge of the DC power line to the transceiver

C. The antenna system

D. FCC rules limiting mobile output power on the 75 meter band

What is one disadvantage of using a shortened mobile antenna as opposed to a full size antenna?

A. Short antennas are more likely to cause distortion of transmitted signals

B. Short antennas can only receive vertically polarized signals

C. Operating bandwidth may be very limited

D. Harmonic radiation may increase

Which of the following may cause receive interference in a radio installed in a vehicle?

A. The battery charging system

B. The fuel delivery system

C. The vehicle control computer

D. All these choices are correct

Richard P. Clem

16 ANTENNAS

One of the most basic antennas is **the half-wave dipole**. As the name implies, it is a half wavelength long, and it is fed in the center. Most commonly, a dipole is run horizontally, but it can be run in any direction. The radiation pattern of a dipole antenna (the direction it "gets out" best) is a a figure-eight at right angles to the antenna.

The formula for calculating the length of a dipole is 468 divided by the frequency, and this gives you the length in feet. For example, if you want a dipole for 14.250 MHz, the length would be 468/14.250, or just over 32 feet.

This is a very useful formula, and you will use it frequently. However, for purposes of the test, you can calculate all of the antenna lengths very easily, as long as you remember the wavelength of the frequency. For example, 14.250 MHz is part of the 20 meter band, so the wavelength is approximately 20 meters. A half wavelength would be approximately 10 meters, so a half-wave dipole would be about 10 meters long. One meter equals about three feet, so this antenna would be approximately 30 feet long, and the closest answer on the test is 32 feet.

Plain-English Study Guide for the FCC Amateur Radio General Class License

The radiation pattern for a horizontal half-wave dipole is generally a figure-eight pattern at right angles to the antenna. However, if this antenna is mounted closer to the ground (less than a half wavelength high), then the pattern is almost omnidirectional.

Also, as the half-wave dipole is lowered (when it gets below 1/4 wave above ground), then the antenna's impedance decreases steadily.

Normally, the half-wave dipole is fed in the center. If the feedpoint is moved toward the ends, then the antenna's impedance increases.

One advantage of a horizontal antenna (such as a horizontal dipole) is that it has lower ground reflection losses.

The other basic antenna is the **quarter-wave vertical**. Again, as the name implies, the length of this antenna is approximately one quarter of the wavelength. For example, one question on the test asks for the length of a quarter-wave vertical for 28.5 MHz. You can use the formula 234/28.5 to get the length in feet, which gives a result of about 8 feet. But again, for all of the questions on the test, you can get an answer that is close enough simply by looking at the wavelength of the frequency given, and dividing that by 4. Here, 28.5 MHz is part of the 10 meter band, so the wavelength is about 10 meters. Therefore, a quarter wavelength antenna would be about 2.5 meters long, or about 7.5 feet. Once again, if you look at the possible answers, the closest one is 8 feet, which is the right answer.

A quarter-wave vertical generally requires ground radials underneath it. Even if the antenna is mounted at a high location, these radials are still used, and form an artificial ground for the antenna. Very often, these radials slope downward slightly. The reason for this is that sloping the radials downward causes the antenna's impedance to increase slightly. Bringing them downward increases the impedance to approximately 50 ohms, which is a better match for most amateur equipment.

A vertical antenna gets out equally well in all directions. The test uses a fancy term for this, namely that the antenna is omnidirectional in azimuth.

If the quarter-wave vertical antenna is mounted on the ground, then the radial wires should be mounted on the surface or buried a few inches below the ground.

Some hams can get surprisingly good results by using nothing more than an end-fed random wire antenna. However, such an antenna has disadvantages. One of them is that you may experience RF burns when touching metal objects in your station. For the test, you also need to know that an end-fed half-wave antenna has a very high impedance.

Generally, an antenna will have a greater bandwidth if the elements have a larger diameter. For a wire antenna, this effect is almost negligible, and the diameter of the wire doesn't noticeably affect bandwidth. However, for larger antennas made of aluminum tubing, the effect can be noticeable.

Many amateurs use directional antennas of some kind. These include the **Yagi**, the **log periodic**, and the **cubical quad**. Obviously, a directional antenna is the best to use for minimizing interference, because the signal would only go in one direction.

The Yagi antenna consists of a driven element, and one or more parasitic elements, called **reflectors** or **directors**. The driven element is the only piece with an electrical connection to the radio. It is basically a half-wave dipole. Behind it is a reflector, which is a slightly longer element. As the name implies, it reflects the signal back toward the driven element.

A director is a slightly shorter element in front of the driven element. As the name implies, it directs the radio waves to move in that direction. This is easy to remember, because when you picture a Yagi, it looks like an arrow, pointing in the direction where the signal will be strongest.

The amount by which the strength is increased is called the gain of the antenna. In the case of the Yagi, the antenna has gain in the direction where the arrow is pointing.

Plain-English Study Guide for the FCC Amateur Radio General Class License

Most Yagis used by amateurs have three elements. However, adding additional directors to the front will increase the gain of the antenna. Also, increasing the boom length will increase gain.

The front to back ratio of a Yagi antenna (or any directional antenna) compares the power radiated in the major direction, as compared to the power radiated in the opposite direction.

The **main lobe** of any directional antenna means simply the direction of maximum radiated field strength.

Yagi antennas are used because they help reduce interference from other stations or behind the antenna. (One question asks why this is true on the 20 meter band. That makes it a trick question. This is true of any band.)

The maximum theoretical forward gain of a three element, single-band Yagi antenna is approximately 9.7 dBi. You can also stack Yagi antennas (use more than one). If you stack two 3-element Yagi antennas, this results in an gain increase of about 3 dB. (This is easy to remember, because from the chapter on decibels, you know that 3 dB is the same as doubling a signal. It stands to reason that if you have two antennas, you'll have twice as much gain.)

There are many variables at work with a Yagi. The physical length of the boom, the number of elements, and the spacing of the elements, all can be adjusted to optimize forward gain, front-to-back ratio, or SWR bandwidth.

Many Yagi antennas are fed with a **gamma match**, which is used to match the relatively low feed-point impedance to the 50 ohms used by most amateur equipment. Also, when a Yagi is fed with a gamma match, it is not necessary to insulate the elements from the boom.

Another type of match used for a Yagi antenna is a **beta** or **hairpin** match. This is a shortened transmission line stub placed at the feedpoint for impedance matchin.

Two or more Yagi antennas can be **stacked**, one above the other, about a half wavelength apart. If so, the gain is about 3 dB higher than one antenna alone. (This should be easy to remember. It stands to reason that two antennas will put out twice as much signal. And as you remember, when you double something, that is an increase of 3 dB.) It does this primarily by narrowing the main lobe in elevation.

A quad antenna is similar to a Yagi, but each element is made up of a square loop of wire. The total length of the wire is a full wavelength, so each side is 1/4 wavelength. The forward gain of a two-element quad is about the same as a three-element Yagi.

A quad antenna is horizontally polarized if it is fed in the center of one of the horizontal wires. It is vertically polarized if it is fed at the center of one of the vertical wires.

Like a Yagi, a quad also has a reflector, which is slightly larger (about 5% larger) than the driven element. Therefore, each side is slightly more than 1/4 wavelength.

The gain of a two-element **delta loop** is also about the same as a three-element Yagi. As the name implies, each element of a delta loop looks like the Greek letter delta, or a triangle. Therefore, each leg of the triangle is about 1/3 wavelength long.

A log periodic is another type of directional antenna. It's main advantage is a wide bandwidth. It gets its name from the fact that the length and spacing of the elements increase logarithmically from one end to the other.

When you have any kind of directional antenna, you need to figure out how to point it in the right direction—toward the other station. One good way of doing this is with an **azimuthal projection map**. This is a map of the world centered on a particular location, namely, your location. This makes it easy to see at a glance the direction toward a station on another continent. Occasionally, you will want to work a station over the long path. This means that instead of using the short way around the world, your signal will travel the opposite direction and take the long way around the world. To do this, you would point your antenna in the opposite direction, or 180 degrees from its short-path heading. For example, if the station was to your northwest, then the long-path heading would be to the southeast.

Antenna **traps** are used in multiband antennas. They make the antenna longer on some frequencies. But at higher frequencies, the signal is "trapped" and cannot use the longer portion of antenna beyond the trap. One disadvantage of any type of multiband antenna is that they have poor harmonic rejection.

A **beverage** antenna is a very long and low directional receiving antenna. It is often used for receiving on the lower HF bands. It is not used for transmitting because it has very high losses.

A **halo** antenna is a horizontal antenna sometimes used for portable work on VHF or UHF. It is omnidirectional. A **screwdriver** antenna is often used for mobile HF work. It has an inductor (coil) at the bottom that is changed to tune the antenna. The motor from an electric screwdriver is sometimes used to make the adjustment.

There are some questions on the test about horizontal loop antennas. You need to know that a small loop has nulls broadside to the loop. You also need to know that a large loop (multi-wavelength) has a pattern that is omnidirectional, but with a low vertical angle of radiation.

What is one disadvantage of a directly fed random-wire HF antenna?

A. It must be longer than 1 wavelength

B. You may experience RF burns when touching metal objects in your station

C. It produces only vertically polarized radiation

D. It is more effective on the lower HF bands than on the higher bands

Which of the following is a common way to adjust the feed point impedance of a quarter wave ground plane vertical antenna to be approximately 50 ohms?

A. Slope the radials upward

B. Slope the radials downward

C. Lengthen the radials

D. Shorten the radials

Which of the following best describes the radiation pattern of a quarter-wave, ground-plane vertical antenna?

A. Bi-directional in azimuth

B. Isotropic

C. Hemispherical

D. Omnidirectional in azimuth

What is the feed-point impedance of an end-fed half-wave antenna?

A. Very low

B. Approximately 50 ohms

C. Approximately 300 ohms

D. Very high

How does antenna height affect the horizontal (azimuthal) radiation pattern of a horizontal dipole HF antenna?

A. If the antenna is too high, the pattern becomes unpredictable

B. Antenna height has no effect on the pattern

C. If the antenna is less than 1/2 wavelength high, the azimuthal pattern is almost omnidirectional

D. If the antenna is less than 1/2 wavelength high, radiation off the ends of the wire is eliminated

Where should the radial wires of a ground-mounted vertical antenna system be placed?

A. As high as possible above the ground

B. Parallel to the antenna element

C. On the surface of the Earth or buried a few inches below the ground

D. At the center of the antenna

How does the feed-point impedance of a 1/2 wave dipole antenna change as the antenna is lowered from 1/4 wave above ground?

A. It steadily increases

B. It steadily decreases

C. It peaks at about 1/8 wavelength above ground

D. It is unaffected by the height above ground

How does the feed-point impedance of a 1/2 wave dipole change as the feed-point location is moved from the center toward the ends?

A. It steadily increases

B. It steadily decreases

C. It peaks at about 1/8 wavelength from the end

D. It is unaffected by the location of the feed point

Which of the following is an advantage of a horizontally polarized as compared to vertically polarized HF antenna?

A. Lower ground reflection losses

B. Lower feed-point impedance

C. Shorter Radials

D. Lower radiation resistance

What is the approximate length for a 1/2-wave dipole antenna cut for 14.250 MHz?

A. 8 feet

B. 16 feet

C. 24 feet

D. 33 feet

Plain-English Study Guide for the FCC Amateur Radio General Class License

What is the approximate length for a 1/2-wave dipole antenna cut for 3.550 MHz?

A. 42 feet

B. 84 feet

C. 132 feet

D. 263 feet

What is the approximate length for a 1/4-wave vertical antenna cut for 28.5 MHz?

A. 8 feet

B. 11 feet

C. 16 feet

D. 21 feet

Which of the following would increase the bandwidth of a Yagi antenna?

A. Larger diameter elements

B. Closer element spacing

C. Loading coils in series with the element

D. Tapered-diameter elements

What is the approximate length of the driven element of a Yagi antenna?

A. 1/4 wavelength

B. 1/2 wavelength

C. 3/4 wavelength

D. 1 wavelength

How do the lengths of a three-element Yagi reflector and director compare to that of the driven element?

A. The reflector is longer, and the director is shorter

B. The reflector is shorter, and the director is longer

C. They are all the same length

D. Relative length depends on the frequency of operation

How does the gain of two three-element, horizontally polarized Yagi antennas spaced vertically 1/2 wavelength apart typically compare to the gain of a single three-element Yagi?

A. Approximately 1.5 dB higher

B. Approximately 3 dB higher

C. Approximately 6 dB higher

D. Approximately 9 dB higher

What is a beta or hairpin match?

A. It is a shorted transmission line stub placed at the feed point of a Yagi antenna to provide impedance matching

B. It is a ¼ wavelength section of 75 ohm coax in series with the feed point of a Yagi to provide impedance matching

C. It is a series capacitor selected to cancel the inductive reactance of a folded dipole antenna

D. It is a section of 300 ohm twinlead used to match a folded dipole antenna

What configuration of the loops of a two-element quad antenna must be used for the antenna to operate as a beam antenna, assuming one of the elements is used as a reflector?

A. The driven element must be fed with a balun transformer

B. There must be an open circuit in the driven element at the point opposite the feed point

C. The reflector element must be approximately 5 percent shorter than the driven element

D. The reflector element must be approximately 5 percent longer than the driven element

How does increasing boom length and adding directors affect a Yagi antenna?

A. Gain increases

B. Beamwidth increases

C. Weight decreases

D. Wind load decreases

What does "front-to-back ratio" mean in reference to a Yagi antenna?

A. The number of directors versus the number of reflectors

B. The relative position of the driven element with respect to the reflectors and directors

C. The power radiated in the major radiation lobe compared to the power radiated in exactly the opposite direction

D. The ratio of forward gain to dipole gain

What is meant by the "main lobe" of a directive antenna?

A. The magnitude of the maximum vertical angle of radiation

B. The point of maximum current in a radiating antenna element

C. The maximum voltage standing wave point on a radiating element

D. The direction of maximum radiated field strength from the antenna

Which of the following can be adjusted to optimize forward gain, front-to-back ratio, or SWR bandwidth of a Yagi antenna?

A. The physical length of the boom

B. The number of elements on the boom

C. The spacing of each element along the boom

D. All of these choices are correct

Which of the following is an advantage of using a gamma match for impedance matching of a Yagi antenna to 50-ohm coax feed line?

A. It does not require that the elements be insulated from the boom

B. It does not require any inductors or capacitors

C. It is useful for matching multiband antennas

D. All of these choices are correct

Approximately how long is each side of a quad antenna driven element?

A. 1/4 wavelength

B. 1/2 wavelength

C. 3/4 wavelength

D. 1 wavelength

How does the forward gain of a two-element quad antenna compare to the forward gain of a three-element Yagi antenna?

A. About the same

B. About 2/3 as much

C. About 1.5 times as much

D. About twice as much

What is the primary purpose of antenna traps?

A. To permit multiband operation

B. To notch spurious frequencies

C. To provide balanced feed-point impedance

D. To prevent out of band operation

What is the advantage of vertical stacking of horizontally polarized Yagi antennas?

A. Allows quick selection of vertical or horizontal polarization

B. Allows simultaneous vertical and horizontal polarization

C. Narrows the main lobe in azimuth

D. Narrows the main lobe in elevation

Which of the following is an advantage of a log periodic antenna?

A. Wide bandwidth

B. Higher gain per element than a Yagi antenna

C. Harmonic suppression

D. Polarization diversity

Plain-English Study Guide for the FCC Amateur Radio General Class License

Which of the following describes a log periodic antenna?

A. Element length and spacing vary logarithmically along the boom

B. Impedance varies periodically as a function of frequency

C. Gain varies logarithmically as a function of frequency

D. SWR varies periodically as a function of boom length

What is the primary use of a Beverage antenna?

A. Directional receiving for low HF bands

B. Directional transmitting for low HF bands

C. Portable direction finding at higher HF frequencies

D. Portable direction finding at lower HF frequencies

Which HF antenna would be the best to use for minimizing interference?

A. A quarter-wave vertical antenna

B. An isotropic antenna

C. A directional antenna

D. An omnidirectional antenna

Which of the following is a disadvantage of multiband antennas?

A. They present low impedance on all design frequencies

B. They must be used with an antenna tuner

C. They must be fed with open wire line

D. They have poor harmonic rejection

Which of the following describes an azimuthal projection map?

A. A map that shows accurate land masses

B. A map that shows true bearings and distances from a particular location

C. A map that shows the angle at which an amateur satellite crosses the equator

D. A map that shows the number of degrees longitude that an amateur satellite appears to move westward at the equator with each orbit

How is a directional antenna pointed when making a "long-path" contact with another station?

A. Toward the rising Sun

B. Along the gray line

C. 180 degrees from its short-path heading

D. Toward the north

Plain-English Study Guide for the FCC Amateur Radio General Class License

What is the radiation pattern of a dipole antenna in free space in the plane of the conductor?

A. It is a figure-eight at right angles to the antenna

B. It is a figure-eight off both ends of the antenna

C. It is a circle (equal radiation in all directions)

D. It has a pair of lobes on one side of the antenna and a single lobe on the other side

In which direction is the maximum radiation from a portable VHF/UHF "halo" antenna?

A. Broadside to the plane of the halo

B. Opposite the feed point

C. Omnidirectional in the plane of the halo

D. Toward the halo's supporting mast

How does a "screwdriver" mobile antenna adjust its feed-point impedance?

A. By varying its body capacitance

B. By varying the base loading inductance

C. By extending and retracting the whip

D. By deploying a capacitance hat

In which direction or directions does an electrically small loop (less than 1/3 wavelength in circumference) have nulls in its radiation pattern?

A. In the plane of the loop

B. Broadside to the loop

C. Broadside and in the plane of the loop

D. Electrically small loops are omnidirectional

What is the combined vertical and horizontal polarization pattern of a multi-wavelength, horizontal loop antenna?

A. A figure-eight, similar to a dipole

B. Four major loops with deep nulls

C. Virtually omnidirectional with a lower peak vertical radiation angle than a dipole

D. Radiation maximum is straight up

Plain-English Study Guide for the FCC Amateur Radio General Class License

17 FEED LINES

In previous chapters, we have discussed transmitters and antennas. This chapter discusses the issues involved with hooking them together.

First of all, one device is sometimes used to match the transmitter's output (which usually has an impedance of 50 ohms) to an antenna that might have a different impedance. On the test, that device is called an **antenna coupler** or **matching network**, although it might also be called an antenna tuner or a transmatch. If a matching network is connected at the transmitter, this can lower the standing wave ratio (SWR) as "seen" by the transmitter. But the SWR on the feed line will remain the same as it was previously.

Ideally, everything should have a matching impedance: The transmitter, the feed line, and the antenna. When you buy feed line, it will have a specified characteristic impedance of a certain number of ohms. This means that it will match with an antenna having that same impedance. If the cable's characteristic impedance is not matched to the feed-point impedance of the antenna, this will result in the presence of standing waves on the antenna feed line. This is worded in a slightly different way in one of the questions: If there is reflected power, this is caused by a difference between feed-line impedance and antenna feed-point impedance.

Richard P. Clem

Most coaxial cable used by amateurs has a characteristic impedance of either 50 or 75 ohms. Flat "window line" transmission line normally has an impedance of 450 ohms. The characteristic impedance of a piece of cable is determined by its physical dimensions. Only a parallel conductor feed line is covered on the test. In that case, the characteristic impedance is determined by the distance between the centers of the conductors, and the radius of the conductors.

The **standing wave ratio (SWR)** is the measurement of how well an antenna, feed line, and transmitter are matched. Most amateurs own an SWR meter for making this measurement. A slightly more sophisticated instrument for making this measurement is a **directional wattmeter**. Either of these is connected between the transmitter and antenna, and a reading is taken while transmitting to the antenna.

One reason to avoid a high SWR is because if the transmission line has a high loss to start with, the high SWR will increase the loss. A lossy transmission line will also hide the fact that the SWR is high, because the SWR will read artificially low.

An **antenna analyzer** can also be used to measure SWR. To make this measurement, the antenna analyzer is connected to the antenna and feed line (but not to the transmitter). One thing to remember when using an antenna analyzer to make this measurement is that a strong signal from a nearby transmitter may affect the accuracy of the measurements. An antenna analyzer can also be used to determine the impedance of an unknown or unmarked coaxial cable.

SWR can also be calculated, and there are a number of questions on the test that ask you to calculate it. Fortunately, all of these questions are quite simple. The question gives two impedance figures, in ohms, and asks you to calculate the SWR. In every question on the test, you can calculate the SWR simply by dividing the larger value by the smaller number. For example, one question asks the SWR resulting from a 50-ohm feed line and a 200-ohm load. The answer is 200 divided by 50, or 4. SWR is

Plain-English Study Guide for the FCC Amateur Radio General Class License

always written as a ratio, with the larger number first, and the second number as 1. So in this case, the correct answer is 4:1. When you say this out loud, you say the SWR is "four to one". This simple rule works for all of the SWR calculations on the test. Just divide the larger number by the smaller number, and write that number followed by " :1".

One thing to remember about coaxial cable is that it can attenuate the signal more than twin lead. The attenuation becomes much greater at higher frequencies.

What type of device is often used to match transmitter output impedance to an impedance not equal to 50 ohms?

A. Balanced modulator

B. SWR Bridge

C. Antenna coupler or antenna tuner

D. Q Multiplier

Which of the following can be determined with a directional wattmeter?

A. Standing wave ratio

B. Antenna front-to-back ratio

C. RF interference

D. Radio wave propagation

Which of the following must be connected to an antenna analyzer when it is being used for SWR measurements?

A. Receiver

B. Transmitter

C. Antenna and feed line

D. All of these choices are correct

What is the interaction between high standing wave ratio (SWR) and transmission line loss?

A. There is no interaction between transmission line loss and SWR

B. If a transmission line is lossy, high SWR will increase the loss

C. High SWR makes it difficult to measure transmission line loss

D. High SWR reduces the relative effect of transmission line loss

What is the effect of transmission line loss on SWR measured at the input to the line?

A. The higher the transmission line loss, the more the SWR will read artificially low

B. The higher the transmission line loss, the more the SWR will read artificially high

C. The higher the transmission line loss, the more accurate the SWR measurement will be

D. Transmission line loss does not affect the SWR measurement

Plain-English Study Guide for the FCC Amateur Radio General Class License

What problem can occur when making measurements on an antenna system with an antenna analyzer?

A. SWR readings may be incorrect if the antenna is too close to the Earth

B. Strong signals from nearby transmitters can affect the accuracy of measurements

C. The analyzer can be damaged if measurements outside the ham bands are attempted

D. Connecting the analyzer to an antenna can cause it to absorb harmonics

What is a use for an antenna analyzer other than measuring the SWR of an antenna system?

A. Measuring the front-to-back ratio of an antenna

B. Measuring the turns ratio of a power transformer

C. Determining the impedance of coaxial cable

D. Determining the gain of a directional antenna

Which of the following factors determine the characteristic impedance of a parallel conductor antenna feed line?

A. The distance between the centers of the conductors and the radius of the conductors

B. The distance between the centers of the conductors and the length of the line

C. The radius of the conductors and the frequency of the signal

D. The frequency of the signal and the length of the line

What are the typical characteristic impedances of coaxial cables used for antenna feed lines at amateur stations?

A. 25 and 30 ohms

B. 50 and 75 ohms

C. 80 and 100 ohms

D. 500 and 750 ohms

What is the typical characteristic impedance of "window line" parallel transmission line?

A. 50 ohms

B. 75 ohms

C. 100 ohms

D. 450 ohms

What might cause reflected power at the point where a feed line connects to an antenna?

A. Operating an antenna at its resonant frequency

B. Using more transmitter power than the antenna can handle

C. A difference between feed line impedance and antenna feed point impedance

D. Feeding the antenna with unbalanced feed line

Plain-English Study Guide for the FCC Amateur Radio General Class License

How does the attenuation of coaxial cable change as the frequency of the signal it is carrying increases?

A. Attenuation is independent of frequency

B. Attenuation increases

C. Attenuation decreases

D. Attenuation reaches a maximum at approximately 18 MHz

What must be done to prevent standing waves on an antenna feed line?

A. The antenna feed point must be at DC ground potential

B. The feed line must be cut to an odd number of electrical quarter wavelengths long

C. The feed line must be cut to an even number of physical half wavelengths long

D. The antenna feed-point impedance must be matched to the characteristic impedance of the feed line

If the SWR on an antenna feed line is 5 to 1, and a matching network at the transmitter end of the feed line is adjusted to 1 to 1 SWR, what is the resulting SWR on the feed line?

A. 1 to 1

B. 5 to 1

C. Between 1 to 1 and 5 to 1 depending on the characteristic impedance of the line

D. Between 1 to 1 and 5 to 1 depending on the reflected power at the transmitter

What standing wave ratio will result from the connection of a 50-ohm feed line to a non-reactive load having a 200-ohm impedance?

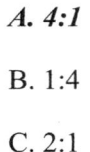

A. 4:1

B. 1:4

C. 2:1

D. 1:2

What standing wave ratio will result from the connection of a 50-ohm feed line to a non-reactive load having a 10-ohm impedance?

A. 2:1

B. 50:1

C. 1:5

D. 5:1

What standing wave ratio will result from the connection of a 50-ohm feed line to a non-reactive load having a 50-ohm impedance?

A. 2:1

B. 1:1

C. 50:50

D. 0:0

Plain-English Study Guide for the FCC Amateur Radio General Class License

18 TESTS AND TEST EQUIPMENT

The exam has some questions about common pieces of test equipment.

One piece of equipment covered by the test is the **oscilloscope**, which is used to view various waveforms. It has an advantage over a voltmeter in that complex waveforms can be measured. The oscilloscope shows its inputs as a graph on the screen. It shows one value vertically, and the other value horizontally. To accomplish this, it contains both horizontal and vertical channel amplifiers. An oscilloscope can be used to check the RF envelope of a transmitted signal. In that case, the vertical input is attached to the attenuated RF output of the transmitter. (You're not asked this on the test, but the horizontal input in this case would be generated internally by the oscilloscope, and would show time.)

Another common piece of test equipment is the **voltmeter**. A good voltmeter has a high input impedance, because a high input impedance decreases the loading on the circuit being measured. Voltmeters come in both analog and digital versions. The digital voltmeter shows the number of volts in a digital display. An analog voltmeter has a mechanical meter movement. A digital voltmeter usually has better precision for most uses. However, analog meters also have advantages. They are useful when you need to make an adjustment, such as adjusting a tuned circuit. With an

Richard P. Clem

analog meter, it's easier to see whether the value is going up or going down.

There are a couple of questions on the test about the two-tone test, which analyzes the linearity of a transmitter. When doing a two-tone test, it is important that the two tones are not harmonically related.

What item of test equipment contains horizontal and vertical channel amplifiers?

A. An ohmmeter

B. A signal generator

C. An ammeter

D. An oscilloscope

Which of the following is an advantage of an oscilloscope versus a digital voltmeter?

A. An oscilloscope uses less power

B. Complex impedances can be easily measured

C. Input impedance is much lower

D. Complex waveforms can be measured

What signal source is connected to the vertical input of an oscilloscope when checking the RF envelope pattern of a transmitted signal?

A. The local oscillator of the transmitter

B. An external RF oscillator

C. The transmitter balanced mixer output

D. The attenuated RF output of the transmitter

Why is high input impedance desirable for a voltmeter?

A. It improves the frequency response

B. It decreases battery consumption in the meter

C. It improves the resolution of the readings

D. It decreases the loading on circuits being measured

What is an advantage of a digital voltmeter as compared to an analog voltmeter?

A. Better for measuring computer circuits

B. Better for RF measurements

C. Better precision for most uses

D. Faster response

What is an instance in which the use of an instrument with analog readout may be preferred over an instrument with a numerical digital readout?

A. When testing logic circuits

B. When high precision is desired

C. When measuring the frequency of an oscillator

D. When adjusting tuned circuits

What type of transmitter performance does a two-tone test analyze?

A. Linearity

B. Carrier and undesired sideband suppression

C. Percentage of frequency modulation

D. Percentage of carrier phase shift

What signals are used to conduct a two-tone test?

A. Two audio signals of the same frequency shifted 90-degrees

B. Two non-harmonically related audio signals

C. Two swept frequency tones

D. Two audio frequency range square wave signals of equal amplitude

Plain-English Study Guide for the FCC Amateur Radio General Class License

19 INTERFERENCE AND GROUNDING

There are a number of questions on the test about dealing with interference with consumer electronics.

One type of interference which is covered on the test is caused by arcing at a poor electrical connection. This would cause interference covering a wide range of frequencies.

One type of interference is caused by an audio device being overloaded by a strong radio signal. For example, if a strong SSB signal were overloading a telephone or other audio device, the interference would sound like distorted speech. A CW transmitter would cause interference that sounds like on-and-off humming or clicking. In audio-frequency devices, one way to reduce RF interference might be to use a bypass capacitor. You would use a capacitor with a value that would allow audio signals to pass through but block radio-frequency signals.

When you took the Technician exam, you had some questions about filters. For the General test, you need to know a few more facts. The **passband** of a filter just means the band of frequencies that are allowed to pass through—in other words, the signal that you don't want to filter out. This bandwidth is usually measured between the "upper and lower half-power frequencies." These are also called the **cutoff frequencies**. Even though you don't want to filter out the passband, it will get reduced by inserting a filter. This is called the **insertion loss**. **Ultimate rejection** means a filter's maximum ability to reject signals outside the passband.

ou also need to know that the impedance of a low-pass filter should be about the same as the transmission line into which it is inserted.

The test also includes a number of questions about grounding. These are grouped in with the interference questions, because improper grounding is sometimes the cause of interference. Poor grounding could also cause a painful RF burn when you touch your equipment, if the ground wire has a high impedance on the frequency that you are using. And a resonant ground connection could cause high RF voltages on the enclosures of your station equipment. The best way to avoid stray RF energy in your station is to connect all of the equipment grounds together. Doing so avoids a ground loop, which can cause RF hot spots. One symptom of a ground loop in your station would be reports of "hum" on your transmitted signal. A related type of interference is caused by common-mode current in audio cables. This can be reduced by placing a ferrite bead around the cable.

Intermodulation is a type of interference most common in FM. For the test, you need to know that it is caused by the combining of two signals in a non-linear circuit.

Which of the following might be useful in reducing RF interference to audio-frequency devices?

A. Bypass inductor

B. Bypass capacitor

C. Forward-biased diode

D. Reverse-biased diode

Which of the following could be a cause of interference covering a wide range of frequencies?

A. Not using a balun or line isolator to feed balanced antennas

B. Lack of rectification of the transmitter's signal in power conductors

C. Arcing at a poor electrical connection

D. The use of horizontal rather than vertical antennas

What sound is heard from an audio device or telephone if there is interference from a nearby single-sideband phone transmitter?

A. A steady hum whenever the transmitter is on the air

B. On-and-off humming or clicking

C. Distorted speech

D. Clearly audible speech

What is the effect on an audio device or telephone system if there is interference from a nearby CW transmitter?

A. On-and-off humming or clicking

B. A CW signal at a nearly pure audio frequency

C. A chirpy CW signal

D. Severely distorted audio

What might be the problem if you receive an RF burn when touching your equipment while transmitting on an HF band, assuming the equipment is connected to a ground rod?

A. Flat braid rather than round wire has been used for the ground wire

B. Insulated wire has been used for the ground wire

C. The ground rod is resonant

D. The ground wire has high impedance on that frequency

What effect can be caused by a resonant ground connection?

A. Overheating of ground straps

B. Corrosion of the ground rod

C. High RF voltages on the enclosures of station equipment

D. A ground loop

What technique helps to minimize RF "hot spots" in an amateur station?

A. Building all equipment in a metal enclosure

B. Using surge suppressor power outlets

C. Bonding all equipment enclosures together

D. Low-pass filters on all feed lines

What process combines two signals in a non-linear circuit or connection to produce unwanted spurious outputs?

Plain-English Study Guide for the FCC Amateur Radio General Class License

A. Intermodulation

B. Heterodyning

C. Detection

D. Rolloff

How can a ground loop be avoided?

A. Connect all ground conductors in series

B. Connect the AC neutral conductor to the ground wire

C. Avoid using lock washers and star washers when making ground connections

D. Connect all ground conductors to a single point

What could be a symptom of a ground loop somewhere in your station?

A. You receive reports of "hum" on your station's transmitted signal

B. The SWR reading for one or more antennas is suddenly very high

C. An item of station equipment starts to draw excessive amounts of current

D. You receive reports of harmonic interference from your station

Which of the following would reduce RF interference caused by common-mode current on an audio cable?

A. Placing a ferrite choke around the cable

B. Adding series capacitors to the conductors

C. Adding shunt inductors to the conductors

D. Adding an additional insulating jacket to the cable

What should be the impedance of a low-pass filter as compared to the impedance of the transmission line into which it is inserted?

A. Substantially higher

B. About the same

C. Substantially lower

D. Twice the transmission line impedance

What is the frequency above which a low-pass filter's output power is less than half the input power?

A. Notch frequency

B. Neper frequency

C. Cutoff frequency

D. Rolloff frequency

What term specifies a filter's maximum ability to reject signals outside its passband?

A. Notch depth

B. Rolloff

C. Insertion loss

D. Ultimate rejection

The bandwidth of a band-pass filter is measured between what two frequencies?

A. Upper and lower half-power

B. Cutoff and rolloff

C. Pole and zero

D. Image and harmonic

What term specifies a filter's attenuation inside its passband?

A. Insertion loss

B. Return loss

C. Q

D. Ultimate rejection

Richard P. Clem

20 BATTERIES AND OTHER POWER SOURCES

The test has a number of questions about various batteries and other power sources.

One type of rechargeable battery that is commonly used in portable amateur equipment is the **nickel-cadmium (NiCd)** battery. NiCd batteries have a low internal resistance, which means that they have a high discharge current.

A normal 12 volt car battery is an example of a **lead acid** battery. A lead acid battery should never be discharged below 10.5 volts, because this would reduce the life of the battery.

The exam contains some questions about solar and wind power. The process by which sunlight is changed directly into electricity is called **photovoltaic conversion**. The approximate voltage of a photovoltaic cell is approximately 0.5 volts.

When you hook up a solar panel and a battery, you might connect a diode in series between the two. This prevents the battery from discharging into the panel during darkness.

The question about wind power seems rather obvious: One disadvantage of using wind power for a primary source of power is that you would need

Plain-English Study Guide for the FCC Amateur Radio General Class License

a large storage system (such as a battery) for times when the wind is not blowing.

There are a number of questions on the test about power supplies. A **power supply** is a device that changes normal AC household current to a higher or lower DC voltage to run a piece of equipment. The AC is rectified, and then filtered. The filter network usually consists of capacitors and inductors.

A high-voltage power supply will include **bleeder resistors**. The purpose of these resistors is to discharge the filter capacitors after the power supply is shut off.

In a full-wave bridge power supply, the peak inverse voltage across the rectifiers is equal to the normal peak output voltage of the power supply. In a half-wave power supply, the peak inverse voltage is two times the normal peak output voltage. One way to make a full-wave rectifier circuit is to use two diodes and a center-tapped transformer.

As its name implies, a half-wave rectifier converts only half the AC cycle to DC. Since a full cycle equals 360 degrees, this means that 180 degrees of the AC cycle is converted to DC. The advantage of a half-wave rectifier is that it requires only one diode. A full-wave rectifier converts the full cycle. In other words, it converts the full 360 degrees. The output of an unfiltered full-wave rectifier (connected to a resistive load) is a series of DC pulses at twice the frequency of the AC input.

A **switching power supply** (as opposed to a power supply with a transformer) has the advantage of requiring smaller components.

What is an advantage of the low internal resistance of nickel-cadmium batteries?

A. Long life

B. High discharge current

C. High voltage

D. Rapid recharge

What is the minimum allowable discharge voltage for maximum life of a standard 12 volt lead acid battery?

A. 6 volts

B. 8.5 volts

C. 10.5 volts

D. 12 volts

What is the name of the process by which sunlight is changed directly into electricity?

A. Photovoltaic conversion

B. Photon emission

C. Photosynthesis

D. Photon decomposition

What is the approximate open-circuit voltage from a modern, well-illuminated photovoltaic cell?

A. 0.02 VDC

B. 0.5 VDC

C. 0.2 VDC

D. 1.38 VDC

What is the reason a series diode is connected between a solar panel and a storage battery that is being charged by the panel?

A. The diode serves to regulate the charging voltage to prevent overcharge

B. The diode prevents self discharge of the battery though the panel during times of low or no illumination

C. The diode limits the current flowing from the panel to a safe value

D. The diode greatly increases the efficiency during times of high illumination

Which of the following is a disadvantage of using wind as the primary source of power for an emergency station?

A. The conversion efficiency from mechanical energy to electrical energy is less than 2 percent

B. The voltage and current ratings of such systems are not compatible with amateur equipment

C. A large energy storage system is needed to supply power when the wind is not blowing

D. All of these choices are correct

What useful feature does a power supply bleeder resistor provide?

A. It acts as a fuse for excess voltage

B. It ensures that the filter capacitors are discharged when power is removed

C. It removes shock hazards from the induction coils

D. It eliminates ground loop current

Which of the following components are used in a power-supply filter network?

A. Diodes

B. Transformers and transducers

C. Quartz crystals

D. Capacitors and inductors

What is an advantage of a half-wave rectifier in a power supply?

A. Only one diode is required

B. The ripple frequency is twice that of a full-wave rectifier

C. More current can be drawn from the half-wave rectifier

D. The output voltage is two times the peak output voltage of the transformer

Which type of rectifier circuit uses two diodes and a center-tapped transformer?

A. Full-wave

B. Full-wave bridge

C. Half-wave

D. Synchronous

What portion of the AC cycle is converted to DC by a half-wave rectifier?

A. 90 degrees

B. 180 degrees

C. 270 degrees

D. 360 degrees

What portion of the AC cycle is converted to DC by a full-wave rectifier?

A. 90 degrees

B. 180 degrees

C. 270 degrees

D. 360 degrees

What is the output waveform of an unfiltered full-wave rectifier connected to a resistive load?

A. A series of DC pulses at twice the frequency of the AC input

B. A series of DC pulses at the same frequency as the AC input

C. A sine wave at half the frequency of the AC input

D. A steady DC voltage

Which of the following is an advantage of a switch-mode power supply as compared to a linear power supply?

A. Faster switching time makes higher output voltage possible

B. Fewer circuit components are required

C. High frequency operation allows the use of smaller components

D. All of these choices are correct

Plain-English Study Guide for the FCC Amateur Radio General Class License

21 CONNECTORS AND CONNECTING CABLES

There are a number of questions on the test that require you to identify the names of certain types of connectors. Many of these will be familiar to you, but make sure you memorize the names of any with which you are not familiar. The following list should cover all of these questions:

DE-9: Computer connection used for serial data port.

PL-259: Plug for coaxial cable, commonly used for frequencies up to 150 MHz.

RCA Phono: Plug commonly used for audio signals.

SMA connector: Another type of plug for coaxial cable. It is a small threaded connector suitable for use up to several GHz.

Type N connector: Another type of plug for coaxial cable. It is moisture-resistant and useful up to 10 GHz.

Which of the following connectors would be a good choice for a serial data port?

A. PL-259

B. Type N

C. Type SMA

D. DE-9

Which of these connector types is commonly used for RF connections at frequencies up to 150 MHz?

A. Octal

B. RJ-11

C. PL-259

D. DB-25

Which of these connector types is commonly used for audio signals in Amateur Radio stations?

A. PL-259

B. BNC

C. RCA Phono

D. Type N

Which of the following describes a type-N connector?

A. A moisture-resistant RF connector useful to 10 GHz

B. A small bayonet connector used for data circuits

C. A threaded connector used for hydraulic systems

D. An audio connector used in surround-sound installations

What is a type SMA connector?

A. A large bayonet-type connector usable at power levels more than 1 KW

B. A small threaded connector suitable for signals up to several GHz

C. A connector designed for serial multiple access signals

D. A type of push-on connector intended for high-voltage applications

Richard P. Clem

22 RMS AND PEP

There are a number of questions on the test regarding the various ways to measure AC (or RF) voltage and power. These questions might seem confusing, but if you read the questions carefully and follow the steps explained here, you will be able to answer all of them. On the other hand, if you discover that the math for this part of the test is out of your league, there are only eight questions, and you can simply memorize them.

The **RMS** (root mean square) voltage of an AC (or RF) signal is the value that results in the same power dissipation as a DC voltage of the same value. In other words, if we're worried about power, then RMS is the equivalent DC voltage. If you have a question involving power and AC voltage, then you will need to work with the RMS voltage, which might be different from the voltage value that is given.

The RMS voltage is about 0.7 times the peak voltage. To put it another way, the peak voltage is about 1.4 times the RMS voltage. (Instead of 1.4, it's really the square root of 2, but 1.4 is close enough.)

With this fact in mind, we can answer one of the questions on the test very easily: What is the RMS voltage of a sine wave with a value of 17 volts peak? To get RMS, we multiply peak by 0.7, and we get 17 x 0.7 = 11.9. The closest answer on the test is 12, which is the correct answer.

Plain-English Study Guide for the FCC Amateur Radio General Class License

The next question is a bit of a trick question: What is the **peak-to-peak** voltage of a sine wave that has an RMS voltage of 120 volts? This is something of a trick question because you need to read it very carefully. It is **not** asking for the **peak** voltage. It is asking for the **peak-to-peak** voltage. The peak-to-peak voltage is twice the peak voltage. In other words, the peak-to-peak voltage is the negative peak voltage plus the positive peak voltage.

We start the same as the last problem. We know the RMS voltage is 120, so the peak voltage is 1.4 x 120 = 168. But we need the peak-to-peak voltage, which is twice as much (because it is the 168 volt negative voltage, plus the 168 volt positive voltage). So the answer is 168 + 168 = 336. The closest answer on the test is 339.4, which is the right answer.

The following question is very straightforward: What would be the RMS voltage across a 50-ohm dummy load dissipating 1200 watts. We are given watts and ohms, and we need to find volts. And since we're dealing with power, we need to think in terms of RMS voltage, which is what the question is asking for.

From the first chapter, we remember that watts equals volts squared divided by ohms. So we know that 1200 equals volts squared divided by 50:

$1200 = V^2 \div 50$

We now think back to our high school algebra. (That was the class where you asked the teacher, "but when will we ever have to use this?") We must solve the equation for V:

$1200 = V^2 \div 50$

$1200 \times 50 = V^2$

$60000 = V^2$

Richard P. Clem

Then, we take the square root of both sides of the equation, and get:

244.9 = V

The closest answer is 245, which is the correct answer.

One other question is there to make sure you know the difference between **peak** and **peak-to-peak**. Is asks what the peak-to-peak voltage is if the RMS voltage is 120 volts. First, you find the peak voltage by multiplying 120 x 1.4 for an answer of 168. But we're not done. The question asks for the **peak to peak**, meaning the negative peak plus the positive peak. So we need to multiply the answer by 2, and get 336. The closest answer is 339.4 volts, which is the correct answer.

The next problem is a bit trickier: What is the output PEP from a transmitter if an oscilloscope measures 500 volts peak-to-peak across a 50-ohm resistor connected to the transmitter output? Here, we know volts and ohms, and we need to calculate watts. We know that watts equals volts squared divided by ohms.

But before we use that equation, we need to change the voltage to RMS. Remember: whenever we're working with volts and trying to figure out power, we always need to work with RMS volts.

You need to read the question carefully, because we are starting out with peak-to-peak volts. This is twice the peak voltage. Since the peak-to-peak voltage is 500, the peak voltage is 250.

We know that RMS equals peak times 0.7. So the RMS voltage is 250 x 0.7 = 175.

Now that we have the RMS voltage, we can finally calculate the power. Watts equals volts squared divided by ohms. So here, watts equals 175 squared divided by 50, which equals 612.5. The closest answer is 625 watts, which is the correct answer.

The next problem works the same way: What is the output PEP from a transmitter if an oscilloscope measures 200 volts peak-to-peak across a 50-

Plain-English Study Guide for the FCC Amateur Radio General Class License

ohm dummy load connected to the transmitter output? Again, we have volts and ohms, and need to calculate watts.

We first change the 200 volts peak-to-peak to 100 watts peak. Then, we multiply by 0.7 to get 70 volts RMS. Then, we get the power as being volts squared divided by ohms. When we punch that into our calculator, we get an answer of 98 watts. The closest answer is 100 watts, which is the correct answer.

For the final two questions, you need to know that for an unmodulated carrier, the **peak envelope power (PEP)** is the same as the average power. In other words, since there is no modulation, the amplitude of the signal never changes. And since it remains the same, it never changes from the average value. That fact allows you to answer the following two questions: What is the ratio of peak envelope power to average power for an unmodulated carrier? The answer is 1.00, because the two values are the same.

The other question about this fact is worded differently: What is the output PEP of an unmodulated carrier if an average reading wattmeter connected to the transmitter output indicates 1060 watts? Since it is an unmodulated carrier, the average and the PEP are the same. So the answer is 1060 watts.

What is the output PEP from a transmitter if an oscilloscope measures 200 volts peak-to-peak across a 50-ohm dummy load connected to the transmitter output?

A. 1.4 watts

B. 100 watts

C. 353.5 watts

D. 400 watts

What value of an AC signal produces the same power dissipation in a resistor as a DC voltage of the same value?

A. The peak-to-peak value

B. The peak value

C. The RMS value

D. The reciprocal of the RMS value

What is the RMS voltage of a sine wave with a value of 17 volts peak?

A. 8.5 volts

B. 12 volts

C. 24 volts

D. 34 volts

What is the ratio of peak envelope power to average power for an unmodulated carrier?

A. .707

B. 1.00

C. 1.414

D. 2.00

What would be the RMS voltage across a 50-ohm dummy load dissipating 1200 watts?

A. 173 volts

B. 245 volts

C. 346 volts

D. 692 volts

What is the output PEP of an unmodulated carrier if an average reading wattmeter connected to the transmitter output indicates 1060 watts?

A. 530 watts

B. 1060 watts

C. 1500 watts

D. 2120 watts

What is the output PEP from a transmitter if an oscilloscope measures 500 volts peak-to-peak across a 50-ohm resistor connected to the transmitter output?

A. 8.75 watts

B. 625 watts

C. 2500 watts

D. 5000 watts

What is the peak-to-peak voltage of a sine wave with an RMS voltage of 120.0 volts?

A. 84.8 volts

B. 169.7 volts

C. 240.0 volts

D. *339.4 volts*

23 ELECTRICAL AND RF SAFETY

The test covers a number of areas about electrical and radio frequency (RF) safety.

Radio frequency energy is potentially dangerous to humans. It can injure humans by heating body tissues. Section 97.13 of the FCC rules requires you to take certain steps to ensure compliance with RF safety regulations. There are two possible questions on the test which ask what steps you need to take to ensure compliance. You are required to perform a routine RF exposure evaluation, to see whether your signal exceeds the maximum permissible exposure (MPE). All of the following are ways to perform this evaluation:

1. By calculation based upon FCC OET Bulletin 65

2. By calculation based on computer modeling

3. By measurement of field strength using calibrated equipment.

If you do use the third method (actual measurement), then you need to use both a calibrated field-strength meter and a calibrated antenna. Since most hams do not have this equipment, they usually use one of the other two methods. (A field-strength meter, even though it's not calibrated, can be used for other purposes, such as monitoring relative RF output when making antenna and transmitter adjustments, for determining the radiation pattern of an antenna, or for close-in radio direction finding.)

When you do this evaluation, there are three important factors that you will take into account. These factors are:

1. The duty cycle

2. The frequency

3. The power density.

A signal with a greater duty cycle (in other words, the signal is transmitting power a larger percentage of the time) would have more effect. For example, an FM or RTYY signal is transmitting at full power 100% of the time, and would have a greater effect than an SSB signal, whose power level rises and falls with the transmitted audio. If you are using a mode with a low duty cycle, then you are permitted to have greater exposure levels. For this reason, when you calculate the exposure level, you will usually use the "time averaged" exposure, which means the total RF exposure averaged over a certain time. (It's also important to know the duty cycle of the mode you are using, because a mode with a high duty cycle might exceed your transmitter's average power rating.)

A signal with a higher frequency will have a greater effect than a lower frequency signal. And a signal with a larger power density would have a greater effect.

If you do an evaluation and it shows that the RF energy exceeds the permissible limits, then you must take action to prevent human exposure to the excessive RF fields.

There are a few specific questions on the test about exposure in certain situations. First, if your exposure evaluation shows that a neighbor might receive more than the allowable limit of RF exposure from the main lobe of a directional antenna, then you need to take precautions to make sure that the antenna cannot be pointed in their direction.

Second, you need to be especially careful of maximum permitted exposure when using indoor transmitting antennas. You need to make sure that MPE levels are not exceeded in occupied areas.

Third, when you are working on your own antennas, you should turn off the transmitter and disconnect the feed line.

Fourth, there are special concerns about an antenna mounted on the ground. Because people can come closer to it than an antenna mounted in the air, you need to be especially careful that it's protected against unauthorized access so that no one will be exposed to RF in excess of the maximum limits.

The test also covers some electrical safety issues. This includes the following points about household electrical wiring. These safety rules are covered by the National Electrical Code, which is the final word on issues about electrical safety inside your ham shack.

1. If you are operating a piece of equipment operated by 240 volt single-phase AC power (such as an amplifier), then the hot wires should be protected by fuses or circuit breakers.

2. A 20-amp household circuit needs to be wired with 12 gauge (AWG number 12) wire. That is the minimum size of the wire.

3. A 15-amp household circuit would be wired with 14 gauge (AWG number 14) wire.

The test also asks one question about **Ground Fault Circuit Interrupters (GFCI)**. These, of course, are the resettable outlets found in most kitchens and bathrooms in the United States. A GFCI will disconnect (or "trip", as most people say), when current flows from one or more of the hot wires directly to ground.

Equipment should be grounded for safety. The metal enclosures of your equipment should be attached to ground so that hazardous voltages cannot appear on the chassis.

Your tower should also be grounded for lightning protection. It is very important not to use soldered joints on the wires running from the tower to the ground rods. This is because if lightning strikes, the solder will probably be destroyed by the heat. All lightning protection grounds should be bonded together with all other grounds.

There are three questions on the test about generator safety. If you ever connect your house to an emergency generator, you must first disconnect the incoming utility power feed. If you do not, you could energize the lines outside your house, and possibly injure or kill a lineman.

If you do use an emergency generator, it should obviously be located in a well ventilated area. And obviously, it should not be used inside an occupied area, because of the danger of carbon monoxide poisoning.

Every year, we read about hams who are killed while putting up an antenna. There are two questions about climbing a tower. First, if you are using a safety belt or harness, you always make sure that the belt is rated for your weight and is not past its expiration date.

And if you are climbing a tower that supports electrically powered devices, then you should make sure that supply power to the tower is locked out and tagged.

There is one safety question about soldering. Most solder used for electrical work is lead-tin solder. Because lead can contaminate food, you should always wash your hands carefully after handling solder.

Many transmitters and other equipment with high voltages are equipped with a power supply interlock. This is a switch which automatically shuts off dangerous voltages if the cabinet is opened.

What is one way that RF energy can affect human body tissue?

A. It heats body tissue

B. It causes radiation poisoning

C. It causes the blood count to reach a dangerously low level

D. It cools body tissue

Which of the following properties is important in estimating whether an RF signal exceeds the maximum permissible exposure (MPE)?

A. Its duty cycle

B. Its frequency

C. Its power density

D. All of these choices are correct

How can you determine that your station complies with FCC RF exposure regulations?

A. By calculation based on FCC OET Bulletin 65

B. By calculation based on computer modeling

C. By measurement of field strength using calibrated equipment

D. All of these choices are correct

What does "time averaging" mean in reference to RF radiation exposure?

A. The average time of day when the exposure occurs

B. The average time it takes RF radiation to have any long-term effect on the body

C. The total time of the exposure

D. The total RF exposure averaged over a certain time

What must you do if an evaluation of your station shows RF energy radiated from your station exceeds permissible limits?

A. Take action to prevent human exposure to the excessive RF fields

B. File an Environmental Impact Statement (EIS-97) with the FCC

C. Secure written permission from your neighbors to operate above the controlled MPE limits

D. All of these choices are correct

What effect does transmitter duty cycle have when evaluating RF exposure?

A. A lower transmitter duty cycle permits greater short-term exposure levels

B. A higher transmitter duty cycle permits greater short-term exposure levels

C. Low duty cycle transmitters are exempt from RF exposure evaluation requirements

D. High duty cycle transmitters are exempt from RF exposure requirements

Why is it important to know the duty cycle of the data mode you are using when transmitting?

A. To aid in tuning your transmitter

B. Some modes have high duty cycles which could exceed the transmitter's average power rating.

C. To allow time for the other station to break in during a transmission

D. All of these choices are correct

Which of the following steps must an amateur operator take to ensure compliance with RF safety regulations when transmitter power exceeds levels specified in part 97.13?

A. Post a copy of FCC Part 97 in the station

B. Post a copy of OET Bulletin 65 in the station

C. Perform a routine RF exposure evaluation

D. All of these choices are correct

What type of instrument can be used to accurately measure an RF field?

A. A receiver with an S meter

B. A calibrated field strength meter with a calibrated antenna

C. An SWR meter with a peak-reading function

D. An oscilloscope with a high-stability crystal marker generator

Which of the following instruments may be used to monitor relative RF output when making antenna and transmitter adjustments?

A. A field-strength meter

B. An antenna noise bridge

C. A multimeter

D. A Q meter

Which of the following can be determined with a field strength meter?

A. The radiation resistance of an antenna

B. The radiation pattern of an antenna

C. The presence and amount of phase distortion of a transmitter

D. The presence and amount of amplitude distortion of a transmitter

What is one thing that can be done if evaluation shows that a neighbor might receive more than the allowable limit of RF exposure from the main lobe of a directional antenna?

A. Change to a non-polarized antenna with higher gain

B. Post a warning sign that is clearly visible to the neighbor

C. Use an antenna with a higher front-to-back ratio

D. Take precautions to ensure that the antenna cannot be pointed in their direction

What precaution should you take if you install an indoor transmitting antenna?

A. Locate the antenna close to your operating position to minimize feed line radiation

B. Position the antenna along the edge of a wall to reduce parasitic radiation

C. Make sure that MPE limits are not exceeded in occupied areas

D. Make sure the antenna is properly shielded

What precaution should you take whenever you make adjustments or repairs to an antenna?

A. Ensure that you and the antenna structure are grounded

B. Turn off the transmitter and disconnect the feed line

C. Wear a radiation badge

D. All of these choices are correct

What precaution should be taken when installing a ground-mounted antenna?

A. It should not be installed higher than you can reach

B. It should not be installed in a wet area

C. It should limited to 10 feet in height

D. It should be installed such that it is protected against unauthorized access

Which wire or wires in a four-conductor connection should be attached to fuses or circuit breakers in a device operated from a 240 VAC single phase source?

A. Only the two wires carrying voltage

B. Only the neutral wire

C. Only the ground wire

D. All wires

According the National Electrical Code, what is the minimum wire size that may be used safely for wiring with a 20 ampere circuit breaker?

A. AWG number 20

B. AWG number 16

C. AWG number 12

D. AWG number 8

Which size of fuse or circuit breaker would be appropriate to use with a circuit that uses AWG number 14 wiring?

A. 100 amperes

B. 60 amperes

C. 30 amperes

D. 15 amperes

Which of the following is a primary reason for not placing a gasoline-fueled generator inside an occupied area?

A. Danger of carbon monoxide poisoning

B. Danger of engine over torque

C. Lack of oxygen for adequate combustion

D. Lack of nitrogen for adequate combustion

Which of the following conditions will cause a Ground Fault Circuit Interrupter (GFCI) to disconnect the 120 or 240 Volt AC line power to a device?

A. Current flowing from one or more of the hot wires to the neutral wire

B. Current flowing from one or more of the hot wires directly to ground

C. Over-voltage on the hot wire

D. All of these choices are correct

Why must the metal enclosure of every item of station equipment be grounded?

A. It prevents blowing of fuses in case of an internal short circuit

B. It prevents signal overload

C. It ensures that the neutral wire is grounded

D. It ensures that hazardous voltages cannot appear on the chassis

Which of these choices should be observed when climbing a tower using a safety belt or harness?

A. Never lean back and rely on the belt alone to support your weight

B. Confirm that the belt is rated for the weight of the climber and that it is within its allowable service life

C. Ensure that all heavy tools are securely fastened to the belt D-ring

D. All of these choices are correct

What should be done by any person preparing to climb a tower that supports electrically powered devices?

A. Notify the electric company that a person will be working on the tower

B. Make sure all circuits that supply power to the tower are locked out and tagged

C. Unground the base of the tower

D. All of these choices are correct

Why should soldered joints not be used with the wires that connect the base of a tower to a system of ground rods?

A. The resistance of solder is too high

B. Solder flux will prevent a low conductivity connection

C. Solder has too high a dielectric constant to provide adequate lightning protection

D. A soldered joint will likely be destroyed by the heat of a lightning strike

Which of the following is a danger from lead-tin solder?

A. Lead can contaminate food if hands are not washed carefully after handling the solder

B. High voltages can cause lead-tin solder to disintegrate suddenly

C. Tin in the solder can "cold flow" causing shorts in the circuit

D. RF energy can convert the lead into a poisonous gas

Which of the following is good practice for lightning protection grounds?

A. They must be bonded to all buried water and gas lines

B. Bends in ground wires must be made as close as possible to a right angle

C. Lightning grounds must be connected to all ungrounded wiring

D. They must be bonded together with all other grounds

What is the purpose of a power supply interlock?

A. To prevent unauthorized changes to the circuit that would void the manufacturer's warranty

B. To shut down the unit if it becomes too hot

C. To ensure that dangerous voltages are removed if the cabinet is opened

D. To shut off the power supply if too much voltage is produced

What must you do when powering your house from an emergency generator?

A. Disconnect the incoming utility power feed

B. Insure that the generator is not grounded

C. Insure that all lightning grounds are disconnected

D. All of these choices are correct

Which of the following is covered by the National Electrical Code?

A. Acceptable bandwidth limits

B. Acceptable modulation limits

C. Electrical safety inside the ham shack

D. RF exposure limits of the human body

Which of the following is true of an emergency generator installation?

A. The generator should be located in a well ventilated area

B. The generator should be insulated from ground

C. Fuel should be stored near the generator for rapid refueling in case of an emergency

D. All of these choices are correct

ABOUT THE AUTHOR

Richard P. Clem, WØIS, has been a licensed amateur since 1974, having previously held the calls WNØMEB and WBØMEB. His previous works include the Plain-English Study Guide for the Technician Exam, the novel *Caretaker*, and a multi-band antenna construction article published in *QST*. He is married to the illustrator, Yippy Clem, KCØOIA, and resides in St. Paul, Minnesota. He is an attorney in private practice. You can visit his websites: RichardClem.com or w0is.com, or his blog: OneTubeRadio.com.

Made in the USA
Middletown, DE
21 April 2023